WHEN WE ARE HUMAN

Notes from the Age of Pandemics

Also by **JOHN ZERZAN:**

Elements of Refusal (1988, 1999)
Future Primitive (1994)
Against Civilization (1999, 2004)
Running on Emptiness (2001)
Twilight of the Machines (2008)
Origins (2010)
Future Primitive Revisited (2012)
Why Hope? (2015)
A People's History of Civilization (2018)

John Zerzan

WHEN WE ARE HUMAN

Notes from the Age of Pandemics

feralhouse.com

When We Are Human
Notes from the Age of Pandemics

Copyright 2021 John Zerzan
All Rights reserved.

ISBN: 9781627311120

Cover art by Sehejveer Singh via The Multicultural Children's
Art Museum and Education Center
arts4kidsoregon.org

Design: designSimple

Feral House
1240 W Sims Way #124
Port Townsend WA 98368

Printed in the USA

For Alice
Words fail.

TABLE OF CONTENTS

INTRODUCTION
by James V. Morgan

Relatively few Western humans are willing to acknowledge how bad things really are now, the actual depth of it all, the reality of what is gone; not just the ecology, but the whole package; the social, the psychological, and physical array of what it actually means to be human; what has worked and what has not; the truth about what *Homo sapiens'* evolutionary fitness actually is, and the lies about what it isn't.

John Zerzan is one of the few who unflinchingly tells it how it is. Not in theory. John's writing as a whole offers arguably the most complete package of multifaceted objective reality to inform us what has actually happened to humans since a few of us (who were sociopaths) developed socioeconomic and political complexity a very short time ago. What has happened as a result of the emergence of elite-driven socioeconomic complexity? Rapid-fire evolutionary maladaptation has happened, and the effects accelerate every minute of every day in the 21st-century hyper-techno-domestication world.

Not only do most people have zero answers, most people incessantly avoid coping with the answers. Mind that there is an entire industry called academia supposedly tasked with providing answers (data) that will inevitably solve our human problems. Yet it is such an irony that academia, in so many ways, did develop its own quite accurate map(s) to the maze decades and decades ago. Academia has long acknowledged in so many ways that the complex, stratified, ultra-domesticated mass-society pathway is a catastrophic one, but the experts from the academies have simultaneously snake-oil-peddled their postmodern excuses to the masses, ultimately just to protect their own skins, as any threat to domesticated life means also a solidified threat to their own lives of complete domesticated dependency.

Never mind the world, or the integrity of our species, let me just log this new journal article in my CV so I can pick up the next research grant and get tenured with six digits. Meanwhile, yes, the world is fucked and most humans have fallen into utter madness. Read between the lines of whatever latest scientific publication you can find and you'll see it's all there. The scientific experts do certainly say we are fucked. But they also love postmodern excuses.

Ever since John got me tuned in to what postmodernism really is nearly two decades ago, identifying PM rationalizations and excuses has become daily fare for me. Nearly every person one might attempt speaking to about the reality of our situation seems to have mastered the PM gaslight as a response. It's their day-to-day self-preservation: abandoning the death ship means giving up on all of this, abandoning my domesticated mental and physical 'Life Ship'; so at all costs Do Not Abandon. Easier to chirp snide PM remarks and go back to poking away at my (Poison) Apple.

Did you ever consider what that logo on your little toy means? The Programmers obviously have known what they are ultimately reaching toward.

For the good of all, it's time for all willing accomplices to start reading and listening to what John Zerzan has to say. John has got things figured out on a level that most people just don't, even the so-called experts—academic, activist, anarchist or whoever else. This is not just about reading some literature. Take JZ's analysis and figure out how to apply the critical points to your life and your future plans. Even better, take JZ's analysis of our predicament and figure out how to apply it to raising your children.

As a 'professional' academic myself, my take is that in the future the Zerzanophiles will maybe have inherited the earth and things will be light-years better than they are now. There is Hope. Abandon the death ship and step onto the authentic Life Ship. It's still there waiting for you to get on board. The gangplank is right before your eyes. First step: trash that stupid Bite-of-the-Poison-Apple-Phone and start learning how to be a human being again. What to do from there will become increasingly obvious with every step. It's When We Are Human.

The wide-ranging essays that comprise this book are so many points of light in a kaleidoscope you won't likely find elsewhere. Taken together—or separately for that matter—they illumine basic realities and may just help in these dark days.

PRE-HISTORY

Systems-wide failure everywhere one looks. What ISN'T failing? Every civilization so far has collapsed. Now there is only this global civilization and it is FAILING.

The perils and pathologies of modern life have come as a surprise to many. These pitfalls didn't show up overnight. The current reality—or unreality—can't be understood without some grasp of how it began, what drives it.

Umair Haque (*Eudaimonia*, July 3, 2020) wrote, "If Life Feels Bleak, It's Because Our Civilization is Beginning to Collapse." The coronavirus is one of many warnings that we can now see the end of civilization.

How did we get to this terminal place?

These essays try to shed some light on how it happened, and what's at stake.

NEWS FROM PREHISTORY: AN UPDATE

Symbolic culture, the defining feature of modern humans, is quite recent, while non-symbolic culture—and intelligence—go back very much farther. About 30,000 years for the former, three million years in terms of the latter. I've addressed this before, most recently in "The Way We Used to Be,"[1] and the following is largely an extension or update of that essay.

Contra Henry de Lumley, the symbolic is not "one of the essential dimensions of human cognition."[2] We are the only human species to symbolize, and yet cognition certainly extends to our very, very distant forebears. We are symbolic animals, living within layers of symbolic representations where nothing is allowed to be merely itself. This conceit defines reality in countless ways. Consciousness, for example, can only take place within the symbolic. Erich Neumann sees the origin of consciousness in myth, to cite one baseless example.[3]

Communication cannot be properly said to take place unless it is symbolic. Michael Haworth has explored "Telepathy and Intersubjectivity in Derrida, Husserl and Levinas,"[4] and Freud had no trouble assuming that early humans were telepathic.[5] The cognition that enables expertise is not usually reliant on the symbolic, including

---

1   "The Way We Used to Be," in John Zerzan, *Future Primitive Revisited* (Port Townsend, WA: Feral House, 2012), pp. 110–124.

2   "The Emergence of Symbolic Thought," Colin Renfrew and Iain Morley, eds., *Becoming Human: Innovation in Prehistoric Material and Spiritual Culture* (New York: Cambridge University Press, 2009), p. 10.

3   Erich Neumann, *The Origins and History of Consciousness* (Princeton, NJ: Princeton University Press, 1970).

4   Michael Haworth, "Telepathy and Intersubjectivity in Derrida, Husserl and Levinas," *The Journal of the British Society for Phenomenology*, 45:3 (2014).

5   Sigmund Freud, *New Introductory Lectures on Psychoanalysis* (New York: Norton, 1933), p. 55.

language. We are slowly discovering more about the richness of pre-symbolic culture, including ever-earlier examples of Paleolithic intelligence.

Culture in the widest sense is far from solely possessed by humans. A fine reminder is *The Cultural Lives of Whales and Dolphins* by Hal Whitehead and Luke Rendell,[6] about cetaceans who think, feel, and live communally in a web of culture developed about 30 million years ago.

Concerning our own family tree, in the beginning there were the hominin species (e.g., *Ardipithecus*, *Australopithecus*) and the *Homo* species. We were fully bipedal this side of six million years ago, but not yet "human." A fairly recent *Ardipithecus ramidus* find is a fossilized skeleton nicknamed "Ardi" who lived about 4.4 million years ago; a more famous ancestor is "Lucy" from 3.4 million years ago. Much debate continues as to the earliest appearance of humans (e.g., *Homo erectus*, *Homo habilis*).[7]

In March 2015 Kaye Reed of Arizona State University and her colleagues reported finding the oldest *Homo* fossil, dating back 2.8 million years, found in Ethiopia.[8] In June of the same year there was another Ethiopian find, half of a jawbone, dated from 3.3 to 3.5 million years. The latest evidence fuels the hominin vs. *Homo* discussion, but also raises questions as to the adequacy of those distinctions. "It makes us stop and rethink everything," said American paleoanthropologist Carol V. Ward.[9]

---

6   Hal Whitehead and Luke Rendell, *The Cultural Lives of Whales and Dolphins* (Chicago: University of Chicago Press, 2014).

7   John Gurche, *Shaping Humans* (New Haven: Yale University Press, 2013), p. 126. Also Fred Spoor, "Paleoanthropology: The Middle Pliocene Gets Crowded," *Nature* 521 (27 May 2015).

8   Villmoare, Brian, et al., "Early Homo at 2.8 Ma from Ledi-Geraru, Ethiopia," *Science* 347 (12 June 2015).

9   Carl Zimmer, "Bones to Pick on Evolution," *New York Times*, June 2, 2015.

The fact that some extremely old fossilized remains have distinctly human features (e.g., shape of hands or feet, arm length)[10] only deepens the confusion, but the extent of cognitive capacities is a question of still greater significance.

Analysis of stone tools found near Lake Turkana, Kenya in 2011 verifies that they are 3.3 million years old, some 700,000 years earlier than those previously known.[11] The earliest previous evidence of tool-making, also from east Africa, was dated 2.6 m.y.a. A similar, supportive find is that of bones from before 3.39 m.y.a. "that show unambiguous stone-tool cut marks for flesh removal and percussion marks for marrow access."[12] The fashioning of even the simplest of stone tools is a feat of mind not exhibited by non-human primates even when trained by humans.[13] Much of what we know is extrapolated from the evidence of stone tools; they are artifacts that endure. There was likely a wealth of other activity whose traces have disappeared, e.g., woodworking, bone and antler tools, cordage from similar periods. A 2019 finding in southern Israel included 283 small precision tools used for butchering an elephant, dating from some 500,000 years ago.

We know that *Homo erectus* managed repeated sea crossings to the Indonesian island of Flores, a distance of at least 20 kilometers.[14] The discovery of 500,000-year-old stone-tipped

---

10 Russell H. Tuttle, *Apes and Human Evolution* (Cambridge, MA: Harvard University Press, 2014), p. 248.

11 John Noble Willford, "Stone Tools from Kenya Are Oldest Yet Discovered," *New York Times*, May 20, 2015.

12 S.P. McPherron et al., "Evidence for Stone-Tool-Assisted Consumption of Animal Tissues Before 3.39 mya at Dikika, Ethiopia," Nature 466 (12 August 2011).

13 Gurche, *op. cit.*, p. 124.

14 Robert G. Bednarik, "Replicating the First Known Sea Travel by Humans: the Lower Pleistocene Crossing of the Lombok Strait," *Journal of Human Evolution*, 16:3 (2001).

spears in South Africa upset the long-standing opinion that such hafting was unknown before 300,000 years ago.[15]

The evidence record shows a clear pattern of developed abilities at ever-earlier times. Other recent findings support this, including a *Journal of Human Evolution* article that focuses on cooking at around 1.9 million years ago.[16] It discusses scavenged meat, arguing that *Homo erectus* would not have emerged without cooking. Eating carrion, which clearly took place at least this early, would not have been safe unless the meat was cooked. Ewen Calloway looks at 1.5 m.y.a. human footprints in Kenya as evidence of an early antelope hunt.[17] A September 2015 sensation was the announcement of a new species, *Homo naledi*, found in South Africa and dating from 2.5 to 2.8 m.y.a., of unusually modern appearance and possibly practicing deliberate burial vastly earlier than any known symbolic activity.[18]

We were beings who lived in direct touch with this Earth while avoiding the virus of symbolic pseudo-life, domestication, and civilization—but not for want of intelligence. Our species is unique, mainly in a negative sense, having brought ruin and estrangement to every corner of the world.

Women as Paleolithic tool-makers[19] brings to mind another dimension of hunter-gatherer band society. A 2011

---

15 Jayne Wilkins et al., "Stone-tipped Spears Lethal, May Indicate Early Cognitive and Society Skills," *Science News*, August 27, 2014.

16 Alex R. Smith et al., "The Significance of Cooking for Early Hominid Scavenging," *Journal of Human Evolution* 84 (July 2015).

17 Ewen Calloway, "*Homo erectus* Footprints Hint at Ancient Hunting Party," *Nature*, 17 April 2015.

18 Paul H G M Dirks et al., "Geological and taphonomic context for the new hominin species *Homo naledi* from the Dinaledi Chamber, South Aftica." Lee R. Berger et al., "*Homo naledi*, a new species of the genus *Homo* from the Dinaledi Chamber, South Africa." Both articles in *eLife*, September 10, 2015.

19 Nyree Findlay, "Gender and Lithic Studies in Prehistoric Archaeology," Diane Bolger, ed., *A Companion to Gender Prehistory* (Malden, MA: Wiley-Blackwell, 2013).

study of 32 hunter-gatherer groups overturned an earlier assumption that such groups were composed mainly of people who were genetically related. Anthropologists Mark Dyble and Andrea Migliano found that most of them were not related, and that the level of non-relatedness increased with the level of gender equality in the band. They attributed the well-known band features of egalitarianism and cooperation to the conscious influence of women,[20] a powerful reply to those who have characterized references to hunter-gatherer gender equality as an illusory modern/ Romantic/leftist projection.

I think pre-domesticated life may remain an intriguing mystery in many, if not most respects. The perspectives it has already revealed, however, may be of profound importance in the always-worsening straits where Progress places us.

---

20 M. Dyble et al., "Sex Equality Can Explain the Unique Social Structure of Hunter-Gatherer Bands," *Science* 348, 7 June 2015.

WHEN WE WERE HUMAN

When did modern *Homo sapiens* show up? That is, how long have there been people like us? The answer has changed dramatically in recent years, with highly interesting implications.

The long-prevailing consensus was that *Homo* became modern about 40,000 years ago, in the Upper Paleolithic, around the time of the European cave paintings.[1] Wow, has this judgment been radically revised. In 1998 paleo-anthropologist Bernard Campbell found that we were modern 100,000 years ago.[2] 2002 saw John Noble Wilford claim that we were modern "by at least 130,000 years ago."[3] Robert Foley, in 1995, had already put the date as "certainly as far back as 110,000 years ago, and possibly as old as 140,000 years."[4] *Homo sapiens* is 150,000 years old according to Kenneth J. Guest, as of 2014.[5] In 2017, Tibayrenc and Ayala set the date at 200,000 years.[6]

The direction of this revision, and the rapid shift involved, are starkly clear. Galway-Witham and Stringer's "How Did *Homo sapiens* Evolve?" (2015) refers to "evidence for *Homo*

1   Colin Renfrew and Iain Morley, eds., *Becoming Human: Innovation in Prehistoric Material and Spiritual Culture* (New York: Cambridge University Press, 2009), Introduction, p. 1. Paul Mellors and Chris Stringer, *The Human Revolution* (Princeton, NJ: Princeton University Press, 1989).

2   Bernard Campbell, *Human Evolution* (New York: Aldine & Gruyter, 1998), p. 434.

3   John Noble Wilford, "When Humans Became Human," *New York Times*, February 26, 2002.

4   Robert Foley, *Humans Before Humanity* (Cambridge, MA: Blackwell Publishers, 1995), p. 124.

5   Kenneth J. Guest, *Cultural Anthropology* (New York: W.W. Norton, 2014), p. 14.

6   Michael Tibayrenc and Francisco J. Ayala, *On Human Nature* (New York: Academic Press, 2017), p. 3. Concurring is Mark Maslin, *The Cradle of Humanity* (New York: Oxford University Press, 2017), p. 173.

sapiens in Morocco as early as 300,000 years ago."[7] In fact, in 2003 P.S.C. Tacon had already contributed "Behaviourally Modern at 300,000 Before Present: Was My Ancestor Brighter than Yours?"[8]

Precisely what the term "modern" involves/includes probably varies among the anthropologists and archaeologists just cited, but an overall updating and reassessment has arrived. Grant McCall argues that patterns of residential or home base activity in the Lower Paleolithic, as well as shared foraging and hunting practices, are the same as those of modern hunter-gatherers.[9] Home base development and use of fire by circa 400,000 years ago has led Nicholas Roland to a similar conclusion, based on evidence from China.[10] New fossil discoveries have overturned conventional thought about early *Homo* capacities, according to Leslie Aiello and Susan Anton.[11]

A key explanation of the depth of early *Homo* "humanness" is "The Revolution that Wasn't," by Sally McBrearty and Alison Brooks (2000).[12] They argued that the cognitive abilities of early members of our species were indistinguishable from our own. In a 2013 follow-up essay, they presented further research supporting the idea of a "cognitive unity"

7   Julia Galway-Witham and Chris Stringer, "How did *Homo sapiens* evolve?" *Science*, 22 June 2018, pp. 1296–1298.

8   P.S.C. Tacon, "Behaviourally Modern at 300,000 B.P.: Was My Ancestor Brighter than Yours?" Modern Human Origins Conference, Sydney, 30 September 2003.

9   Grant S. McCall, *Before Modern Humans* (Walnut Creek, CA: Left Coast Press, 2015), e.g., pp. 25, 325, 328.

10   Nicholas Rolland, "Was the Emergence of Home Bases and Domestic Fire a Punctuated Event?" *Asian Perspectives* 43:2 (Fall 2004), pp. 248–280.

11   Leslie C. Aiello and Susan C. Anton, "Human Biology and the Origins of Homo," *Current Anthropology* vol. 53, Supplement 6, December 2012.

12   Sally McBrearty and Alison Brooks, "The Revolution that Wasn't," *Journal of Human Evolution* 39:5 (November 2000).

throughout members of the *Homo sapiens* species.[13] John J. Shea's work, for example *"Homo sapiens* Is as *Homo sapiens* Was,"* is founded on the same premise, as the title makes clear.[14]

A new paradigm has emerged.

Looking back much further, we were walking upright more than four million years ago. 3.6-million-year-old footprints found in east Africa show two people walking together with a modern gait.[15] Until recently, the earliest known intentionally modified stone tools (from an Ethiopian site) were dated at 2.6 million years ago. But a 2015 discovery in Kenya has pushed that date back to 3.3 million years ago—a major find.[16]

In 2019, Justin Pargeter and John Shea provided the first extensive overview of prehistoric tool miniaturization, a practice that goes back to extremely early lithic toolmaking.[17] That is, to at least 2.6 million years ago. These are often stunning creations, finely crafted tools less than half an inch long. They were used for cutting, piercing, scraping, etc. It becomes increasingly easy to grasp that we possessed significant capabilities far earlier than the lack of social and material complexity might imply. For this reason, Gowlett,

---

13 Sally McBrearty and Alison Brooks, "Advances in the Study of the Origin of Humanness," *Journal of Anthropological Research* 69:1 (Spring 2013).

14 John J. Shea, *"Homo sapiens* Is as *Homo sapiens* Was," *Current Anthropology* 52:1 (February 2011). John J. Shea, "Refuting a Myth about Human Origins," *American Scientist* 99:2 (March-April 2011).

15 Eric H. Cline, *Those Stones Make a Wall: The Story of Archaeology* (Princeton, NJ: Princeton University Press, 2017), p. 101. Bruce Bower, "African Hominid Fossils Show Ancient Steps Toward a Two-Legged Stride," *Science News*, February 22, 2019.

16 R. Holloway, "Evolution of the Human Brain," in *Handbook of Human Symbolic Evolution* (London: Blackwell, 1999). Michael Balter, "World's Oldest Stone Tools Discovered in Kenya," *Science*, April 14, 2015.

17 Justin Pargeter and John J. Shea, "Going Big versus Going Little: Lithic Miniaturization in Hominin Technology," *Evolutionary Anthropology*, 29 March 2019.

Gamble and Dunbar have argued that "there is at least a 2-million-year social record that must be explored."[18]

From at least 1.5 million years ago, fire was a key development.[19] Hunting of small animals (e.g., rabbits by 400,000 years ago) and larger game (goats and deer) began much earlier than previously thought, according to 2019 *Science Advances* research.[20] The cognition required in stone knapping has long been understood as not substantially different from our own today. [21]

What stands out most to me is the absence of symbolic functioning among these early people. Recent findings underline impressive human capacities at earlier and earlier stages, but with no evidence of symbolic activity, much less of symbolic culture.

Civilization has made the symbolic the measure of intelligence and even of consciousness. Human capacities at remarkably remote times render this notion utterly ridiculous. There was a time when communication wasn't about trading symbols, when the symbolic dimensions of art, number and time awareness did not exist. Robert Bednarik has addressed "Concept-Mediation Marking in the Lower Paleolithic,"[22] regarding very early intelligence in a non-symbolic world.

18 John Gowlett, Clive Gamble and Robin Dunbar, "Human Evolution and the Archaeology of the Social Brain," *Current Anthropology* 53:6 (December 2012).

19 Frances D. Burton: *Fire: The Spark that Ignited Human Evolution* (Albuquerque, NM: University of New Mexico Press, 2011).

20 Bruce Barton, "Hominids May Have Hunted Rabbits as far Back as 400,000 Years Ago," *Science News*, March 6, 2019. Clive Finlayson, *The Improbable Primate: How Water Shaped Human Evolution* (New York: Oxford University Press, 1014, p. 50.

21 Thomas Wynn, *The Evolution of Spatial Competence* (Urbana, IL: University of Illinois Press, 1989).

22 Robert G. Bednarik, "Concept-Mediated Marking in the Lower Paleolithic," *Current Anthropology* 36:4 (August-October 1995).

To me, what we now know of our very distant past leads to the question of the very nature of symbolism's reign over the planet. How to somehow get outside of representation, the symbolic, is a challenge that has been lurking—if not directly posed—for some time. Edmund Husserl's "to the things themselves" comes to mind, the search for a way to be before/beneath the merely conceptual.

A recent entry in the effort toward the direct and unmediated goes by the name of Thing Theory, kicked off, at least in part, by Bill Brown's 2004 cultural studies book *Things*.[23] Cognitive archaeologist Lambros Malafouris turned this emphasis into what he calls Material Engagement Theory.[24] His outlook foregrounds the role of things in the processes of human cognition, stressing the active collaboration between individual and material. As he puts it, with emphasis, "to think through things, in action, without the need of mental representation."[25]

We may be getting closer to directly challenging—and indicting—symbolic culture, whose advent and emergence became viral with domestication and civilization. The realm of estrangement and ruin, in every sphere. Each step into the symbolic has moved us toward alienation and destruction, as we now can more clearly see.

---

23 Bill Brown, ed., *Things* (Chicago: University of Chicago Press, 2004).

24 Lambros Malafournis, *How Things Shape the Mind: A Theory of Material Engagement* (Cambridge, MA: MIT Press, 2013).

25 *Ibid.*, p. 237.

HUMAN NATURE

It's just human nature to _____. Women are by nature _____.

Fill in the blanks. Unexamined essentialism, usually in service to the dominant order.

But a blanket condemnation of essentialism, applied to everything, is its own error. Domestication, for instance, has an essential, core quality: control. It grows broader and deeper, according to its inner logic, and that is easy to see. An open-and-shut case of essentialism!

Human nature is certainly to be rejected in a generally postmodern, no stable meaning or truth culture. Rousseau found our true nature to be that of pre-civilized freedom. His "noble savage" conception is roundly mocked on all sides. But doesn't anarchism rest on the (essentialist) notion that at base, humans are good? And that, as per Rousseau, the problem is that we have been debased and corrupted by various institutions?

Freud's *Civilization and its Discontents* portrayed domestication as an incurable wound to our nature, an unending source of pain that represses our original condition of Eros and freedom. Only the end of domestication/civilization, Freud strongly implied, could cure this fundamental unhappiness. Definitely an essentialist perspective.

For more than 99 percent of our two to three million years as *Homo* species, we lived as mobile hunter-gatherers/foragers. How could this be other than foundational?

Edward O. Wilson proclaimed a "predisposition to religious belief...in all probability an ineradicable part of human nature."[1] This is an absurd judgment, given how very recently (about 3,000 years ago) organized religion

---

1   Edward O. Wilson, *On Human Nature* (Cambridge, MA: Harvard University Press, 1978), p. 169.

entered the picture. Much closer to the mark is the effort by Maryanski and Turner "to discover our 'human nature' by looking at the past—the very distant past...."[2] Our past as foragers and hunters is distant in terms of its duration, but is also recent, considering that domestication is barely 10,000 years old.

"We have not lost, and cannot lose, the genuine impulse of living in balance with the world.... it is the inherent possession of everyone," in the words of Paul Shepard.[3] Similarly, Fredy Perlman referred to the constancy of resistance to Leviathan, the death culture that is civilization.

On a very deep level, it is our nature to want what we have lost.

2   Alexandra Maryanski and Jonathan H. Turner, *The Social Cage: Human Nature and the Evolution of Society* (Stanford, CA: Stanford University Press, 1992), p. 4.

3   Paul Shepard, *Nature and Madness* (Athens, GA: University of Georgia Press, 1982), pp. 27, 26.

FIRE

Evenings are my time to stare at the fire, through the glass of our wood stove. With a nightly martini, maybe with an audiobook playing, whether or not a source of warmth is needed. Seeing the ever-changing flames, letting wordless thought come forth—or not. Staring into the mystery that we all share.

A passage from Hermann Hesse's *Demian* says it well: "Gaze into the fire, into the clouds, and as soon as the inner voices begin to speak...surrender to them. Don't ask first whether it's permitted, or would please your teachers or father, or some god. You will ruin yourself if you do that."

When we anarchists gathered around a fire pit for weekly get-togethers, it seemed so satisfyingly appropriate, the perfect space for conviviality and focus. It was the fact of the fire as a centerpiece that made those sessions so special, I think.

In a September 1954 *Scientific American* article, Loren Eiseley wrote that the use of fires was "doubtless one of the earliest" human practices. Timothy Ingalsbee's entry on the topic in Bron Taylor's *Encyclopedia of Religion and Nature* (Vol. 1) concluded, "*Homo sapiens* became human beings with their knowledge, control and use of fire."

The sense of Ingalsbee's statement is valid, but *Homo* species' fire practices greatly predate the appearance of *H. sapiens*. In his 2019 offering *Architecture and Fire*, Stamatis Zografos finds that the first signs of human settlements coincide with the lighting of fire for warmth, safety, and food preparation. "There is an assumption among architectural historians and theorists that the primitive hut was erected around fire."

Recent findings have put fire use at ever earlier beginnings. "Hominid use of Fire in the Lower and Middle Pleistocene" by Steven R. James, et al. (*Current Anthropology*, February

1989), refers to incidences at Chesowanje, Kenya, dated 1.4 million years ago and at Yuanmow, China, 1.7 million years ago. Mobile foragers, without permanent camps and well-defined hearths, left less clear evidence, which suggests that the origins of fire occurred even earlier.

Although sustained archaeological interest in the topic is sparse, the record in terms of Europe is more studied, the picture a more recent one. We know that humans inhabited Europe in glacial conditions from about 400,000 to 200,000 years ago. Fire must have been a required component for life, argue Lawrence Guy Strauss, "On Early Hominid Use of Fire" (*Current Anthropology*, August 1989) and Wil Roebroeks and Paola Villa, "On the Earliest Evidence for Habitual Use of Fire in Europe" (*PNAS*, March 29, 2011).

Returning to more distant epochs, there have been some recent contributions that help clarify the role of fire in the human record. The presence of fire as accidental (e.g., caused by lightning strikes) may be ruled out when evidence of fire is found deep in caves, for example. "Micro-Stratigraphic Evidence of In Situ Fire in the Acheulean Strata of Wonderwerk Cave, South Africa" by Francisco Berna, et al. (*PNAS*, May 15, 2012) examines just this kind of case. Ash samples from strata of bones, plant material, and stone tools reveal repeated contact with fire 1.8 million years ago.

J.A.J. Gowlett cites consistent evidence of burnt materials from 1.5 m.y.a. on, in "The Discovery of Fire by Humans: A Long and Convoluted Process" (*Philosophical Transactions: Biological Sciences*, 5 June 2016). He also points to the increase in brain size fueled by the better quality diet that cooking enabled. The latter point was emphasized in Robert Wrangham's *Catching Fire: How Cooking Made Us Human* (2009); greater food efficiency released more energy for brain growth in the emergence of *Homo erectus* about 2 m.y.a. Maybe even earlier and in a more general sense, our cognitive abilities developed in tandem with the presence of fire. This is the thesis of Frances Burton's *Fire: The Spark that Ignited Human Evolution* (2011). Our forebears looking into the fire, as we do now.

This progression can be viewed negatively, however. Some argue that use of fire was the first instance of domestication, the distant cornerstone of the relatively recent actual domestication; that it set forth domination of nature, culminating in the domestication of life (plants, animals, including ourselves). But domestication means changing the nature of something; the nature of fire is not changed by using it. Fire can be quenched, but it cannot be made cool, made what it is not.

Fire is vibrant, never the same. The aim of domestication is stasis, sameness. Fire always seeks no control. Fire is not property; domesticates are property. Insofar as we are domesticated, we are property. The property of domestication/civilization.

Passion is often described as fire, fiery. Domestication is the basic removal of passion.

Fire wants to grow and is unruly. A firefighter commented, "I'd rather fight a hundred structure fires than a wildfire. With a structure fire you know where your flames are, but in the woods it can move anywhere. It can come up right behind you."

There certainly have been devastating wildfires, and that's getting worse in a warming, drying biosphere. Australia's mega-fire of early 2020 burned over 2,300 square miles. Only tropical rainforests don't burn, and that is subject to change.

Hunter-gatherer burning practices protected the diversity and sustainability of ecosystems. "Burning the Land" by Fulco Scherjon et al. (*Current Anthropology*, June 2015) indicated that "diverse off-site fire use is as old as the regular use of fire." Such ancient and beneficial methods obviously contrast with destructive outbreaks. M. Kat Anderson's *Tending the*

Wild (2005) discusses fairly recent indigenous fire approaches in California and notes "the almost universal belief among California Indian tribes that catastrophic fires were not a regular natural occurrence," but punishment for lack of respect for the land (p. 57).

Anderson's subtitle, "Native American Knowledge and the Management of California's Natural Resources," should not pass unremarked. The "management" of "resources" is exactly the language of domestication. Its ethos of control has meant a landscape of ruin, contrary to the spirit of the primary indigenous dimension/outlook, as I understand it. Native knowledge should not be looked on as so many techniques, to reform a fundamentally malignant orientation and its values.

Various ways to employ fire, but none to domesticate it. Be like fire.

RITUAL

There are words or terms whose meanings are apparent—and yet elusive. Ritual is one of those tricky ones. Jan A.M. Snoeck writes, "Defining the term 'rituals' is a notoriously problematic task."[1] I wouldn't have thought this to be the case.

It's pretty clear that without ritual there would have been neither myth nor religion.[2] But we can also see that some rituals do not have to do with anything transcendent.

Not only is it somewhat rare that anthropologists and other scholars explain what they mean by ritual, it's "unclear when rituals first originated."[3] Despite the latter gap in our knowledge, many have proclaimed ritual a universal among humans. "There are rituals in all human groups," asserts Pascal Boyer.[4] It is "the basic social act" upon which society is founded, according to Roy Rappaport.[5] Roland Delattre continues the chorus with his view that "ritual is present wherever humanity is present,"[6] and Barry Stephenson assures us that it "must have been present at the very beginning of humanity."[7]

James Fernandez provides the key to this ignorance with a word. He speaks of rites being "at the heart of any genuine

1   Jan A.M. Snoeck, "Defining Rituals," in Jens Kreinath, Jan Snoeck and Michael Stausberg, *Theorizing Rituals* (Boston: Leiden, 2008), p. 1.

2   Colin Renfrew, "Introduction," in Colin Renfrew and Iain Morley, *Becoming Human: Innovation in Prehistoric Material and Spiritual Culture* (New York: Cambridge University Press, 2009), p. 8; and R.A. Segal, "Myth and Ritual," in Kreinath et al., *op. cit.*, p.103.

3   Kreinath et al., "Introductory Essay," *op. cit.*, p. xiii.

4   Pascal Boyer, *Religion Explained* (New York: Basic Books, 2001), p. 231.

5   Roy A. Rappaport, "The Obvious Aspects of Ritual," in *Cambridge Anthropology* 2:3 (1974), p. 5.

6   Roland A. Delattre, "Ritual Resourcefulness and Cultural Pluralism," in *Soundings* 61, p. 281.

7   Barry Stephenson, *Ritual* (New York: Oxford University Press, 2016), p. 21.

human life."[8] The key word is, of course, "genuine," and it signifies humans having arrived at symbolic culture. Ritual is basic to the transmutation of nature into culture, in Lévi-Strauss' sense.[9] For all those who seemingly could not imagine humans without ritual, neither could they grasp humanness as outside of, unmediated by symbolic culture. Even though human society and human capabilities existed hundreds of thousands of years before any evidence of the symbolic. And there is ample recent evidence of hunter-gatherers mocking agriculturalist neighbors for their ritual observances, from the likes of Richard Lee, Colin Turnbull, etc.

Ritual is symbolic behavior that is recognizable insofar as it has form. Formal: standardized and repetitive. Stylized and scripted—and in this sense anticipating modern life.

It is invariant, unidirectional, hierarchical, reinforced by repetition.[10] Rituals are always embedded activities; they are rituals in society, they become the routine of religion, and as both ceremonies and society become more complex, the greater is insistence upon accuracy in routine.

There is also a tendency for ritual to become separated in time and space from daily life. Mircea Eliade's central notion is that the primary intention of all major ritual phenomena is the erasure of temporality and return to a primordial past.

The word derives from the Latin *rit*, referring to number, order. A common, cross-cultural feature is the marking of spatial boundaries,[11] the creation of a controlled environment. It is a communal medium that works toward assurance and solidarity when anxiety or alienation has appeared

8   James W. Fernandez, "Rhetorics," in Kreinath et al., *op. cit.*, p. 656.

9   Claude Lévi-Strauss, *The Savage Mind* (London: Weidenfeld and Nicolson, 1966).

10   Don Handelman, "Framing," in Kreinath et al., *op. cit.*, p. 574.

11   Boyer, *op. cit.*, p. 237.

in social life.[12] Ritual comes to promise an outcome that cannot be obtained by other means, and a way to provide legitimacy. Ritualization is a making of rules, prescribing and proscribing certain forms of behavior. Victor Turner described it as a mechanism that "converts the obligatory into the desirable."[13] That is, not naked coercion but a more subtle means of authority.

Most of the many approaches to the meaning of ritual, its origin and role, avoid societal factors. René Girard focused on a supposed original act of violence as the matrix of all ritual. We are doomed to ritual as sacrifice, to check the constant threat of violence.[14] This view parallels that of the Christian original sin thesis that we are basically sinners in need of redemption.

Sigmund Freud saw the repetitiveness of rituals as the expression of compulsive neurosis, related to the omnipresence of the repressed.[15] The individual in civilization must practice repression over and over, as it is never fully completed. Similarly, ritual must be repeated endlessly as an act of repression of anti-civilization impulses.

More recently, ritual has come to be seen in terms of performance and/or play. Representative of the former approach is Gavin Brown's "Theorizing Ritual as Performance."[16] Although Greek drama is said to have originated in ritual, looking at

---

12   Paul Hockings, "On Giving Salt to Buffaloes. Ritual as Communication," in *Ethnology* 7, p. 411.

13   Victor Turner, *Forest of Symbols* (Ithaca, NY: Cornell University Press, 1967), p. 30.

14   René Girard, *Violence and the Sacred* (Baltimore: The Johns Hopkins University Press, 1977), pp. 18–19.

15   Sigmund Freud, "Obsessive Actions and Religious Practices," in *The Standard Edition of the Complete Psychological Works of Sigmund Freud* (London: Hogarth, 1907).

16   Gavin Brown, "Theorizing Ritual as Performance," in *Journal of Ritual Studies* 17 (2008); and Richard Bradley, *Ritual and Domestic Life in Prehistoric Europe* (New York: Routledge, 2005).

ritual in terms of practice and performance is a move away from social context to the aesthetic dimension. *Ritual, Play and Belief, In Evolution and Early Human Societies*[17] is a recent work that emphasizes play. Rituals as repetitive, rule-governed games. Of course, play may be free and unstructured so that the analogy need not apply.

We can't know the thoughts of our ancestors who ritualized—their satisfactions, their doubts and fears. It is so challenging to plumb the meanings of our own consciousness. And the further back in time, the less, no doubt, does ritual resemble our understanding of it today.

There are guesses that already by the Middle Paleolithic, ca. 100,000 years ago, some routines had begun to assume a ritualistic sense or flavor,[18] as social existence became, very, very slowly more complex. Stephen Shennan discusses examples of the link between ritual and inequality in forager or hunter-gatherer societies.[19] Ritual masters, shamans, were the first who came to hold special ritual roles, specialized authority. The division between those who possessed certain powers and those who did not constituted an obvious differential in society.

But hunter-gatherer life was still intact "on the eve of its complete disruption," as Jacques Cauvin referred to the imminence of domestication, the arrival of the Neolithic.[20] Regional elites held their rituals in early Neolithic sites such as Göbekli Tepe, in what is now southeastern Turkey. The

---

17 Colin Renfrew, Iain Morley, Michael Boyd, eds., *Ritual, Play and Belief, in Evolution and Early Human Societies* (New York: Cambridge University Press, 2018).

18 Gregory J. Wightman, *The Origins of Religion in the Paleolithic* (Lanham, MD: Rowman & Littlefield, 2015), p. 13.

19 Stephen Shennan, "Social Inequality and Cultural Traditions," in James Steele, Stephen Shennan, eds., *The Archaeology of Human Ancestry* (New York: Routledge, 1996), pp. 369, 376.

20 Jacques Cauvin, *The Birth of the Gods and the Origins of Agriculture* (New York: Cambridge University Press, 2000), p. 28.

woman and the bull, representing fertility and domestication, were common images in the region. Only where ritual flourished did priesthoods emerge.[21]

As with civilization itself, in ritual there is assurance, safety; departure from it is danger.

Ritual and symbolic culture in general intersect, and are at the core of emergent civilizations, carriers of estrangement and exploitation in an unfree world. The symbol is the building block of ritual, according to Turner, its "smallest unit."[22] Moving away from nature, the formal rules content, reducing it almost to zero.

Felicitas Goodman points out that "in a ritual of any kind, gesture takes precedence over the spoken word."[23] But it is also true that speech and language are interdependent aspects of the same symbolic domain. Lévi-Strauss observed that ritual elements "serve *in loco verbi*; they are substitutes for words."[24]

Further, ritual systems are analogous to language systems. They can be understood in terms of syntactic rules, as Frits Staal describes it.[25] Ritual is a special kind of language, very condensed in its grammar and syntax. In ritual, words and phrases are usually pared back to the level of "sound bites."

The appearance of art, cave art especially, shows fairly clear evidence of ritualistic activity. The caves in question were not inhabited, in fact not habitable. They were chambers or sanctuaries set off from nature and daily living.

21   Wilson D. Wallis, *Religion in Primitive Society* (New York: F.S. Crofts, 1939), p. 75.

22   Turner, *op. cit.*, p.19.

23   Felicitas D. Goodman, *Ecstasy, Ritual, and Alternate Reality* (Bloomington, IN: Indiana University Press, 1988), p. 16.

24   Claude Lévi-Strauss, *The Naked Man* (New York: Harper & Row, 1984), p. 671.

25   Frits Staal, *Rules without Meaning: Ritual, Mantras and the Human Sciences* (Toronto: Toronto Studies in Religion, 1989), p. 88.

Cave art seems to have originated from ritualist experience; Samuel G. Brandon, noting art's expressive nature, puts ritual before art in a continuum.[26]

Theory has been inadequate to the task of fully accounting for the social role of ritual and its contribution to social structure. There is authority in ritual itself, but also in terms of what happens in society. David Kertzer sees ritual at the heart of political activity, indeed as a vehicle for all forms of authority.[27]

Social relations may be changed via ritual in ways that are not clear to participants, like the subtle arrival of greater complexity and division in society. Ritual may disguise the ways in which irksome duties and roles emerge, to make them seem tolerable, if not desirable.

Transition is generally key to ritual operation, a means of navigating and regulating social systems, of meeting a problematic challenge. To unify society and to alleviate anxiety, according to Robert Alan Segal,[28] or "a device for producing social solidarity," as Talcott Parsons put it.[29]

Given the basically conservative nature of ritual in society, non-participation in ceremonies amounts to defecting from an integrative institution. Ritual makes the social structure explicit, favoring cohesion over individual autonomy.

Humphrey and Laidlaw point out that "in ritual you both are and are not the author of your own acts."[30] The same

---

26 Samuel G.F. Brandon, *Man and God in Art and Ritual* (New York: Charles Scribner's Sons, 1975), p. 8.

27 David Kertzer, *Ritual, Politics, and Power* (New Haven, CT: Yale University Press, 1988), passim.

28 Robert Alan Segal, "Making the Myth-Ritualist Theory Scientific," in *Religion* 30 (2000), pp. 259–271.

29 Talcott Parsons, *Action Theory and the Human Condition* (New York: Free Press, 1978), p. 213.

30 Caroline Humphrey, James Laidlaw, eds., *The Archetypal Action of Ritual* (Oxford: Clarendon Press, 1994), p. 99.

could be said for individuals within the division of labor; that most basic institution, the one that drives the rest.

The individual is submerged within the coordinating embrace of both ritual and division of labor. They share basic formal properties; we can see that routinization, regularization, and repetition, for example, have come to be the basis of life in society. A latent and emergent domestication adheres to both. The person is subsumed more fully by the whole. Standing for the national anthem comes to mind.

The contexts of ritual have varied enormously through the ages. Ekila ritual, common to various forest hunter-gatherers in central Africa, establishes menstrual taboos, but at the same time works to generate and maintain gender egalitarianism.[31] At times ritual has played, and is playing, a defensive role against assaults on Native people and places. Anthropologist Roy Rappaport sees a return to the authority of ritual, to help the threatened environment.[32]Nonetheless, there are primary aspects of ritual as a symbolic institution that should not be overlooked. And as Stephen Shennan has noted, the most egalitarian communities have been those most indifferent to symbolic culture.[33]

31 Jerome Lewis, "Woman's Biggest Husband Is the Moon," radicalanthropology group.org, November 20, 2018.

32 Bron Taylor, *Encyclopedia of Religion and Nature II* (New York: Continuum, 2005), p. 1387.

33 Shennan, *op. cit.* in Steele and Shennan, p. 372.

GONE TO CROATAN

A common response to the anti-civilization/primitivist critique or challenge is incredulity at the notion that anyone would take it literally, seriously. Who, after all, would want to "go back" to a more primitive mode of life?

With that in mind, the record of just such "regression" comes to mind. The fact of "gone to Croatan";[1] abandonment of "advanced" ways of life.

Peter Farb introduces this succinctly, referring to "the appeal that Indian societies held for generation after generation of Whites. No sooner did the first Whites arrive in North America than a disproportionate number of them showed they preferred Indian society to their own."[2] Fairly soon after the Virginia colony was established in 1607, scores of male and female colonists had found Indian mates.

This transculturalization or "Indianization" did not work in the other direction. During the 18th century, as the numbers of white defectors increased and most chose not to return, "it became apparent that few Indians had or wished to become white."[3] Farb underlines this: "Whites who had lived for a time with Indians almost never wanted to leave."[4] The openness and acceptance of the indigenous groups, based on an ethos of sharing, was compelling and generally quite missing among the settlers.

---

1   Allegedly carved on a tree at Roanoke colony. Colonists had decamped to join the nearby Croatan natives.

2   Peter Farb, *Man's Rise to Civilization As Shown by Indians of North America from Primeval Times to the Coming of the Industrial State* (New York: E.P. Dutton, 1968), p. 313.

3   Lin Holdridge, "Visual Representation as a Method of Discourse on Captivity, Focused on Cynthia Ann Parker," in Max Carocci and Stephanie Pratt, eds., *Native American Adoption, Captivity and Slavery in Changing Contexts* (New York: Palgrave Macmillan, 2012), p. 182.

4   Farb, *op. cit.*, p. 314.

From Columbus on, the general Native reception to European arrival was hospitable. That they were met by violence and epidemic diseases is widely known.[5] The Puritans in Massachusetts viewed wild country as the environment of evil, and portrayed Native people in terms of cannibalism and rape. But as Lin Holdridge points out, "the actual instances of white women raped by Indians was less of a documented fact than a prurient fantasy of displaced male fears."[6]

For three hundred years, escaped slaves fled to the sanctuary villages of nearby Indians, where they were welcomed and protected. An estimated 80 percent of African Americans in southeast Virginia, for example, have significant Native ancestry.[7]

Capture of colonists by Natives is a "massive historical reality," as Gary Ebersole put it.[8] The custom of integrating whites into tribes was a successful adjustment to the disastrous decline of population and health occasioned by European invasion. Speaking early on, Frances Slocum observed, "When the Indians thus lose all their children, they adopt some new child as their own, and treat it in all respects like their own. This is why they so often carry away the children of white people."[9] Some were later repatriated to white society, forcibly or otherwise, while many others, white and black, remained, their stories mostly lost.

Frances Slocum's account is available to us, along with other Native-positive experiences such as from Enice

---

5   J.C.H. King, *First Peoples, First Contacts: Native Peoples of North America* (Cambridge, MA: Harvard University Press, 1999), pp. 44–45.

6   Holdridge, *op. cit.*, p. 170.

7   Patrick Minges, "Captives or Captivated: Rethinking Encounters in Early Colonial America," in Carocci and Pratt, *op. cit.*, p. 138.

8   Gary L. Ebersole, *Captured by Texts* (Charlottesville, VA: University Press of Virginia, 1995), p. 2.

9   K.Z. Derounian-Stoda and James Arthur Levernier, *The Indian Captivity Narrative, 1550–1900* (New York: Twayne, 1993), p. 90.

Williams, Isiah Mooso, John Bickell, William Filley, John Dunn Hunter, and William Smith. Mary Jemison, "the white woman of the Genesee," remained with her captors and established a new family network. She saw Indian culture as "essentially more innocent and benevolent than white culture, and she blames most of the problems of her adopted culture onto the corrupting influences of Europeans," in particular "spiritous liquors."[10] J. Norman Heard's study points to "an abundance of evidence that many captives quickly accepted the Indians as their own people and came to regard the whites as enemies."[11]

Five years before Jemison's 1758 capture, Benjamin Franklin wrote these reflections:

> When white persons of either sex have been taken prisoners young by the Indians, and lived a while among them, tho ransomed by their Friends, and treated with all imaginable tenderness to prevail with them to stay among the English, yet in a Short time they become disgusted with our manner of life...and take the first good Opportunity of escaping again into the woods, from whence there is no reclaiming them.[12]

Two excellent modern novels help fill out the picture. *A Light in the Forest*, by Conrad Richter,[13] is the story of John Butler, taken by the Tuscarora, a Delaware tribe, when he was four. He became True Son, where "No one stood between

10   Pauline Turner Strong, *Captive Selves, Captivating Others* (Boulder, CO: Westview Press, 1999), p. 84.

11   J. Norman Heard, *White Into Red* (Metuchen, NJ: Scarecrow Press, 1973), p. 2.

12   Benjamin Franklin, Letter to Peter Collinson, May 9, 1753. teachingamericanhistory.org

13   Conrad Richter, *A Light in the Forest* (New York: Alfred A. Knopf, 1972) [1953].

them and life. They took their joy and meat directly from its hand."[14] He renounced a way of life "where men of their own volition constrained themselves with heavy clothing like harness, where men chose to be slaves to their own or others' property and followed empty and desolate lives far from the wild beloved freedom of the Indian."[15]

Paulette Jiles' *News of the World*[16] tells of Johanna Leonberger, captured in Texas at age six. Her new family "could go without food or water or money or shoes or hats and did not care that they had neither mattresses nor chairs nor oil lamps."[17] Recaptured forcibly and returned to her German uncle and aunt by a man who was paid to do so, she very soon escaped, and found a new life within, yet outside white culture.

Cynthia Ann Parker was a real-life person returned to white society, only to escape twice to her Comanche people. Perceived as wild and untamable, a threat to colonial society, kept against her will, she died early of grief.[18] Until the practice was ended in the 1880s, hundreds of white captives had become more or less completely "Indianized."

Hector de Crèvecoeur's 1782 *Letters from an American Farmer* found Natives in tears as they released their "captives." He termed their social bonds "singularly captivating."[19] The top-drawer American historians, however, decried Indian inhumanity and approved of the genocides applied to them. Herman Melville, for his part, opposed this contemptible attitude, and drew predictable abuse. His first novel, *Typee*,

---

14 *Ibid.*, p. 145.

15 *Ibid.*, p. 179.

16 Paulette Jiles, *News of the World* (New York: William Morrow, 2016).

17 *Ibid.*, p. 63.

18 Holdridge, *op. cit.*, e.g. pp. 174–175.

19 Hector de Crèvecoeur, *Letters from an American Farmer* (New York: E.P. Dutton, 1957) [1782], p. 209.

suffered editorial mutilation and was attacked for alleging "that savage [life] is preferable to civilized life."[20] The great popularity of James Fenimore Cooper's Natty Bumpo character and the real-life frontiersman Daniel Boone testified to the draw of living in nature, far from civilization's strictures.[21] *The Scarlet Letter* by Nathaniel Hawthorne depicts Hester Prynne's estrangement and persecution within white society. The vigor and freedom of the wild and the Natives' place in it beckon to her as a mode of salvation.

Henry David Thoreau shared Boone's disgust that "the axe was always destroying the forest,"[22] and famously wrote, "Give me a wildness whose glance no civilization can endure."[23] The Native in fact served as a kind of model: the Boston Tea Party rebels of 1773 adopted Indian dress in their raid, and the Whiskey Rebels of 1794 drew on the Indian as a symbol of courage and resistance. Thoreau's noble savage is noble insofar as his values are at odds with those of civilization.

Although more commonly associated with Rousseau a century later, the term "noble savage" seems to go back to these 1672 lines by John Dryden:

> I am as free as Nature first made man,
> Ere the base laws of servitude began,
> When wild in woods the Noble Savage ran.[24]

---

20   Joshua David Bellin, *The Demon of the Continent* (Philadelphia: University of Pennsylvania Press, 2001), p. 33.

21   Ebersole, *op. cit.*, p. 210.

22   Richard Slotkin, *Regeneration through Violence: The Mythology of the American Frontier, 1600–1860* (Middletown, CT: Wesleyan University Press, 1973), p. 509.

23   *Ibid.*, p. 535.

24   John Dryden, *The Conquest of Granada* (Berkeley, CA: University of California Press, 1978) [1672], Part 1, Act 1, Scene 1.

Joshua David Bellin's *The Demon of the Continent* (2001)[25] demonstrates that the notion has recently fallen on hard times—and rightly so—according to Bellin and others. His book is a classic postmodern treatment, attacking "Noble Savagism" throughout. Simple dichotomies and "static categories" should be rejected in favor of "cultural ambiguity."[26] What might be called the "who can say?" school is the postmodern effort to undermine any revealed facts or truths. Thus Noble Savagism must be abandoned; the reality is, of course, much more messy, ambiguous, unstable—and ultimately, unknown! The language of no exit, no way to make contact with any realm of freedom.

But certainly there were those who found a way of escaping their suffocating lives, at least for a time. Some found freedom, though they had been taken captive, while others who considered themselves free were unwitting captives of civilization. Women were perhaps most captive of all, draftees in a male enterprise.

A local counselor reports that many returning Peace Corps volunteers have found it very hard to re-adjust to modern reality. An echo, of course, of earlier alienation or estrangement. The Promised Land has become, as F. Scott Fitzgerald wrote in *The Great Gatsby*, a depleted "valley of ashes."[27]

---

25 Bellin, *op. cit.*

26 *Ibid.*, p. 22.

27 Referred to severally in F. Scott Fitzgerald, *The Great Gatsby* (New York: Scribners, 2004) [1922].

HISTORY

The struggle to be human within civilization has always been just that: a continuing, hard-fought struggle. The terrain has shifted over the years, and domestication's grip seems to somehow always extend itself and become deeper and more pervasive. And yet resistance never dies; life demands responses to what tries to strangle it.

As the 21st century unfolds, I think it is easier to see how we have failed, how shallow our efforts have often been. The enormity of the crisis today shows that we need to be equal to its depth. What has been won, compared to what has been lost?

The reigning totality stands ever more revealed, less able to withstand basic questioning. Arundhati Roy sees the possibilities of the situation in her essay "The Pandemic as Portal" (*Financial Times*, April 3, 2020). We must rethink the present, imagine a different future. But what should we keep, what must go? Answers require particulars, historical and otherwise.

WEAVING

Weaving is a process of interlacing slender and flexible fibers to make a fabric. People everywhere have used native plants, animal hair, and other sources to devise shelters, fishnets, bowstrings, floor mats, cushions, carrying bags, water bottles, etc., as well as garments. Today machine-made synthetics pollute the earth and threaten species.

Any weaving, even the most elaborate and accurate, can be done with a minimum of mechanical aid. Looms arrived relatively recently. Perhaps more than any other creative form, what is woven and how it is woven is closely related to its social context. Traditionally, textiles are not about "saving" time or developing division of labor, which obeys the clock. Work, art, play are not separated, do not exist as distinct dimensions. Weaving was not done by specialists; it could be done by women, men, or both.

Plato and Aristotle denigrated handcrafts as involving an inferior grade of knowledge. Its "feel" cannot rise to true understanding, in line with those philosophers' general dismissal of sensual and direct experience. In its own way, Kathryn Sullivan Kruger's *Weaving the Word*[1] conforms to their bias. Kruger finds in textiles the origin of the written text, upholding the symbolic over the actual. In fact, "textile" is related to "texture," not text: *texere*, to weave, and *textura*, texture, in Latin. There is an intimacy that is lost to an overall de-skilling, as civilization advances. To put the text, the symbolic, as the desired result is to devalue the texture of the woven. The path of technological progress is the triumph of the symbolic. This involves an increasing

---

1   Kathryn Sullivan Kruger, *Weaving the Word* (Selinsgrove, PA: Susquehanna University Press, 2001).

separation of conception from execution, as Braverman points out.[2]

No one knows when, with very simple tools or none at all, the first cloth was made. Possibly the first needles, found in what is now eastern Europe and Russia, date to almost 26,000 BCE.[3] Bone needles with eyes so small they could only have been used with very finely spun thread have been found from the Magdalenian period of the Upper Paleolithic in Europe, 17,000 to 12,000 years ago.[4] The oldest known fragment of a textile or basket was discovered on clay fragments in the Czech Republic, 27,000 years old.[5]

Botanists have estimated that over two thousand species of plants produce fibers, including at least one thousand in North America alone. The use of fibers gathered from wild cotton dates back much earlier than the introduction of domesticated cotton. In sub-tropic areas, wild cotton often grows as a 15- to 20-foot tree, with a ten-year quality yield.

There is archaeological evidence that baskets were used in California four millennia ago.[6] Three types of woven sandals have been found in the high desert of eastern Oregon, once belonging to hunter-gatherers more than 13,000 years ago. Margaret Susan Mathewson, writing about Native weavers in California, noted that "Indian people view basketry traditions as one of the strongest ties to their past lifeways."[7]

---

2   Harry Braverman, *Labor and Monopoly Capital* (New York: Monthly Review Press, 1975).

3   Mary Schoeser, *World Textiles* (London: Thames & Hudson, 2003), p. 10.

4   Betty Hochberg, *Spin, Span, Spun* (Santa Cruz, CA: B. and B. Hochberg, 1979), p. 32.

5   Brenda Fowler, "Find Suggests Weaving Preceded Settled Life," *New York Times*, May 9, 1995.

6   M. Kat Anderson, *Tending the Wild* (Berkeley: University of California Press, 2005), p. 223.

7   Margaret Susan Mathewson, *The Living Web: Contemporary Expressions of California Indian Basketry*, Ph.D. dissertation (University of California Berkeley, 1998), p. 1.

According to traditional weavers, "Plants must not feel as though they were simply uprooted and taken away. They must be treated as any other valued member of society."[8]

In *The Romance of French Weaving*, Paul Rodier expresses the opposite and inaccurate orientation, in favor of civilization: "And when a loom is waiting the night of savagery is over."[9]

I have long been drawn to the story of the 19th-century English handloom weavers. They were the heart of the Luddite risings during the first two decades of the 1800s.[10] Their struggles personify autonomy and resistance in the modern history of the West. Probably England's biggest single occupational group, there were about a quarter-million weavers in the early 1800s, a number that grew until the 1830s.

Long accustomed to a casual attitude toward work, the weavers of Manchester "never worked six days in a week; numbers not five, nor even four," economist Arthur Young observed in 1770.[11] Displaying "the old artisan craving for independence," as J.F.C. Harrison put it.[12] Such irregular habits were certainly frowned upon by the managers of English society, who would have preferred docility. Handloom weaving was essentially a family affair; the family mode of production was incompatible with, even opposed to,

---

8   *Ibid.*, p. 97.

9   Paul Rodier, *The Romance of French Weaving* (New York: Tudor Publishing Co., 1939), p. 1.

10   My main offering re Luddites and weavers is "Who Killed Ned Ludd?" in John Zerzan, *Elements of Refusal* (Columbia, MO: CAL Press, 1998). Also, "Industrialism and its Discontents" in my *Why Hope: The Stand Against Civilization* (Port Townsend, WA: Feral House, 2015).

11   Duncan Bythell, *The Handloom Weavers* (Cambridge, UK: Cambridge University Press, 1969), p. 132.

12   Geoffrey Timmins, *The Last Shift: The Decline of Handloom Weaving in Nineteenth-Century Lancashire* (New York: Manchester University Press, 1993), p. 153.

productivity and surplus. Neither age- nor gender-specific, weaving—for a time at least—took place only when time and inclination suited, within the family and with quality goods a priority.

Class strife was "erupting with increasing frequency" by 1750, with weavers in the forefront.[13] Food riots were almost constant in the countryside, and it was very difficult to enforce control, especially over the scattered and recalcitrant handloom folks. Significant embezzlement of yarn from their employers was another weapon used by the weavers to great effect.[14] The fact of being dispersed or scattered has often been seen in a negative light, as detrimental to solidarity or joint action. But weavers were in touch during their weekly visits to take product to their employer's workshop, not to mention meeting at the local alehouse.

Herding the populace into factory discipline and obeying the clock was the recipe for the triumph of domination of the workplace. Industrialization was a striking historical development, heralding a much-needed advance in domestication. This proved hard to achieve, involving decades of protracted struggle. Factory enthusiast Andrew Ure, much cited by Karl Marx, conceded that it was nearly impossible to convert persons past the age of puberty into power loomhands.[15] The earliest factory weavers were children and women, seen as less intractable than men. Very soon the battle against the Machine was joined. In 1792, for example, irate handloom weavers burned Grimshaw's factory at Manchester, which had featured the newly invented Cartwright mechanized looms.

---

13 Bythell, *op. cit.*, p. 99.

14 John K. Walton, *Lancashire: A Social History, 1558–1939* (New York: Manchester University Press, 1987), p. 100.

15 Andrew Ure, *The Philosophy of Manufactures* (London: C. Knight, 1835), p. 15.

The widespread Luddite risings between 1800 and 1820 enjoyed enormous social support. With the anti-mills weavers at its heart, the movement probably reached its peak in 1811–1812. Hammers and torches were wielded in great numbers, and the authorities were stymied by the "nearly unanimous sympathy manifested for the Luddites."[16] As the second decade of the century waned, so did the fortunes of the handloom weavers; a profound rear-guard orientation set in. In 1826 there were several days of widespread loom breaking in Lancashire, but industrialization was plainly winning.

The Plug Riots of 1842 consisted of marauding bands who destroyed the plugs of steam boilers, shutting down power. Handloom weavers held on even into the 1850s and beyond. Pressures increased; the decline in living standards was relentless and dramatic. Marx and Engels rooted for the handloom weavers' demise, calling for an end of resistance to proletarianization. The weavers' long retreat raises the question why so many people chose miserable poverty for so long. Their commitment to an independent way of life was evidently very deep-seated. A richness of social bonds was at stake. Like their Irish counterparts, when men finished their other labors they sat and carded wool for the weaver women, all sharing the latest news, singing, telling stories. They chose autonomy, and tried to defend family and community. "Thus passed away," wrote power loom historian Richard Marsden, "a type of industry, picturesque far beyond its successor."[17]

Resistance by weavers didn't begin with the Luddites. In 265 A.D. the Gallic weavers of Arras went on strike against

---

16   Richard K. Fleischman, Jr., *Conditions of Life among the Cotton Workers of Southeastern Lancashire, 1780–1850* (New York: Garland Publishing, Inc., 1985), p. 61.

17   Bythell, *op. cit.*, p. 267.

the occupying Romans, depriving them of their clothing supply. Arras weavers struck again many times, notably in 1578. Weavers were among the most militant workers in medieval times, especially in Flanders, very often resisting new technological developments.[18]

In the 1760s and 1770s in colonial New England, a spontaneous boycott of British textiles meant that the weaving and wearing of homespun cloth appeared, as a weapon against rule from London. "Homespun" was actually finely woven and often colorful cloth, not a dull or crudely made class of fabric.

Gandhi spearheaded reliance on hand spinning and weaving in the 1930s, to deprive Britain of its large textile exports to the Indian subcontinent.

In the American Southwest, indigenous people wove baskets and sandals with a double twining method, without using a loom. This finger-weaving approach is much more flexible; stitch and color may be varied at any point. The earliest inhabitants in the Southwest lived in small mobile groups, disinclined to carry looms with them. They relied on the life around them for all their needs—especially, increasingly, wild cotton for cloth.

With cotton domestication, commencing at about 900 A.D., loom-type devices appeared and became slowly more complex. Complex society means more complex technology.[19] But weaving remains a link to traditional lifeways, a vital connection in preserving reciprocal relations between Native people and their desert homelands.

Reminiscent of the trajectory of the English handloom weavers, Kathy M'Closkey points out that from the 19th

---

18 For example, N.B. Harte and K.G. Ponting, eds., *Cloth and Clothing in Medieval Europe* (London: Heinemann Educational Books, 1983), p. 108; David Nicholas, *Medieval Flanders* (New York: Longman, 1992), p. 277.

19 Lynn S. Teague, *Textiles in Southwestern Prehistory* (Albuquerque: University of New Mexico Press, 1998), pp. 9, 63, 169.

century on, "the more Navajos wove, the poorer they became."[20] Why Navajo women continued to weave under worsening conditions involved resisting technological innovation. They chose to not produce standardized, decontextualized textiles, and placed no value on time-saving "efficiencies." The clash of perspectives is obvious.

Society is woven, as much as anything that weavers have woven, and continue to weave. The connection is a deep one, with many variations. In Himalayan Ladakh, where women do not weave, it is said that "women spin threads that hold together the social fabric of society."[21] The Dorze weavers in Ethiopia, as yet untouched by mass markets and fashion trends, create traditional handwoven textiles for everyday life with very simple implements.[22] Bedouin Arabs, with their legendary generosity and self-reliance, are very mobile and therefore also weave with very basic equipment. In Guatemala, indigenous Mayans have stood against pressures of change and military assaults, armed importantly by traditional weaving.[23] The T'boli people of the Philippines cleave to the old ways and community via weaving, with its spiritual and Native-oriented basis.[24] Maori people share many handweaving techniques with indigenous cultures elsewhere, and exhibit an emphasis on war and hierarchy.[25]

20 Kathy M'Closkey, *Swept Under the Rug: A Hidden History of Navajo Weaving* (Albuquerque: University of New Mexico Press, 2002), p. 246.

21 Monisha Ahmed, *Living Fabric: Weaving Among the Nomads of Ladakh Himalaya* (Trumbull, CT: Weatherhill, 2002), p. 100.

22 Lucie Mathiszig, "The Dorze Weavers of Ethiopia," *Textile*, 24 August 2015, pp. 180–187.

23 Deborah Chandler and Teresa Cordon, *Traditional Weavers of Guatemala: Their Stories, Their Lives* (Loveland, CO: Thrums Books, 2015).

24 Maria Elena P. Paterno, et al., *Dreamweavers* (Makati City, Philippines: The Bookmark, 2001).

25 Teague, *op. cit.*, p. 188.

As Lynn Teague points out concerning some decorative textiles from late prehistoric North America, one can see in their temporal and spatial depictions "the development of hierarchically organized society and specialized ritual functions."[26] In a similar vein, Theodor Adorno could draw out much of the structure and tensions of society by examining the formal elements of its music.

There is nothing magic or transcendent to be found in the histories of weaving, no radical kernel to serve as a triumphant weapon against domination. But in our deskilled and disempowered condition, there is certainly much to learn from weavers about autonomy and resistance. This cursory survey introduces some of the implications of their struggles. From their material inspiration we can learn some of what we need for a primitive future.[27]

---

26 Julie Paama-Pengelly, *Maori Art and Design: Weaving, Painting, Carving and Architecture* (Auckland, NZ: New Holland, 1010), e.g., p. 18.

27 For more on a "primitive" future, see John Zerzan, *Future Primitive Revisited* (Port Townsend, WA: Feral House, 2012).

ENCLOSED

From early domestication in England, agriculture was carried on under the open-field arrangement. The land was a shared resource in the communal control of small-scale pastoral farmers—a world in which villagers lived their own lives and cultivated the soil independently; originally for subsistence, not for market. Self-governing village communities, from the late Middle Ages up to about 1800, in which "neither county justices nor central government" interfered with daily life.[1]

English peasants possessed an intimate and necessary knowledge of their environment, and in their autonomy could be said to be at least semi-free. Folk traditions and customary rights were the foundations of this status. But over the course of four centuries, the ways of independence and mutuality were attacked and fatally undermined. Anti-communal forces rose up against the commons and its commoners. In Gilbert Slater's words, "The central fact in the history of any English village since the Middle Ages, is expressed in the word 'enclosure'."[2]

The process of privatizing communally held land began slowly, rather inconspicuously, in the 1400s. In this sense the enclosure movement resembles the onset of division of labor or specialization, which at times, especially in its early stages, is steady but gradual. Enclosures went largely unrecorded, although it is also true that there were many important and dramatic instances of resistance.

"The fabric of society was being disrupted," according to Karl Polanyi, and although the very worst instances of enclosure "happened only in patches, the black spots threatened

---

1   W.E. Tate, *The Enclosure Movement* (New York: Walker and Company, 1967), p. 31.

2   Gilbert Slater, *The English Peasantry and the Enclosure of Common Fields* (New York: Augustus M. Kelley, 1968 [1901], p. 1.

to melt into a uniform catastrophe."[3] And no common was ever brought back, just as factories weren't forced to disappear. The loss of communal rights was a profound degradation. Marx famously traced the origins of Britain's urban proletariat to farmers forced off the land by enclosure.[4] This was a major advance in the domestication of humans. Those who managed to remain in agriculture became far poorer, as rents and prices rose steadily. The government's General Report of 1808, for example, contained nearly unanimous complaints from poor people who had lost their cows and could no longer provide milk for their children.[5] To be able to keep a cow was literally a measure of a family's ability to keep starvation at bay.

Enclosure made a few people rich at the expense of the very many. "Civilisation, in this and other guises, was rapidly painting the green spaces black."[6] Which brings to mind the toll on the rest of life. Deforestation had eliminated about 95 percent of original English woodlands by 1700; it was left to enclosure to privatize and destroy the last four million acres.[7] Kirkpatrick Sale adds that "It is almost never reckoned what the cost to the nonhuman species of the sweeping enclosure movement must have been."[8]

Along with the goals of property, profits, and production, another pro-enclosure motive was the need to impose social discipline. People with a strong measure of independence

3   Karl Polanyi, *The Great Transformation: The Political and Economic Origins of Our Time* (Boston: Beacon Press, 2001), p. 35.

4   Karl Marx, *Capital* (London: Swan Sonnenschein & Co., 1901), e.g., p. 769.

5   K.D.M. Snell, *Annals of the Labouring Poor* (Cambridge, U.K.: Cambridge University Press, 1985), p. 174.

6   Polanyi, *op. cit.*, p. 45.

7   Kirkpatrick Sale, *Rebels Against the Future* (Reading, Massachusetts: Addison-Wesley, 1995), p. 54.

8   *Ibid.*, p. 55.

and self-sufficiency can be difficult to oppress or exploit. Like the hand-loom weavers, yeoman smallholders were seen as dangerous and undisciplined. By the late 1600s, polemicists promoted enclosure, denouncing the commons as "seminaries of a lazy, thieving sort of people."[9] In his *Report on Shropshire* (1784), a Mr. Bishton clearly stated a basic goal of enclosure: "that subordination of the lower ranks of society which in the present time is so much wanted, would be thereby considerably secured."[10] Matthew Johnson's *An Archaeology of Capitalism* (1996) discusses a larger context of the ideology of order and control, cultural efforts toward "closure" drive, manifested in areas such as domestic architecture and church liturgy.[11]Alexander Pope's favored Palladian architecture, and classical garden design come to mind: the village common as an enclosed, landscaped park.

Government involvement proved decisive, in the form of parliamentary enclosure awards beginning in the mid-18th century. "Seven million acres were covered by parliamentary awards between 1760 and 1815."[12] Enclosure acts by the State, authored by big landowners, marked a turning point in class robbery. The abolition of the old communal organization of agriculture was greatly accelerated, and the extinction of village community was at stake. More and more, the result was fewer, larger-scale capitalist farms, a shrinking number of semi-independent "cottagers," and a growing host of dependent, landless laborers.

---

9 George Macaulay Trevelyan, *English Social History* (New York: D. McKay, 1962), p. 300.

10 J.L. and Barbara Hammond, *The Village Labourer*, Vol. II (London: Guild Books, 1948 [1911]), p. 31.

11 Matthew Johnson, *An Archaeology of Capitalism* (Wiley, 1996).

12 Peter Mathias, *The First Industrial Nation: The Economic History of Britain 1700–1914* (New York: Routledge, 2001), p. 73.

Enclosure and usury had been stigmatized before the early 16th century; attitudes were shifting toward both land consolidation and capitalist gains. The logic of early capitalist rationality was delayed by deep-rooted customary institutions, restraints on production and productivity. But Jean Calvin, among others, assisted a sea change in Europe's economic culture.

The very meaning of property was in question. Heretofore, landlords ruled, but did not own the land. In the course of the 17th century, clarification and assertion of ownership meant an increasingly absolute definition of property. The countryside became mere property, with the creation of consolidated, individually managed farms. The notion of absolute property in land, subduing the earth at a new level of control, also required subduing the poor.

There were contrary voices raised against these new developments. J. Howlett of Great Dunmow in Essex opposed enclosure fundamentally in 1808: "Let us no longer boast of our improvements, but let us return to our primitive barbarity, and let our flocks and herds resume the undisturbed possession of the forests."[13] In the early 1800s, the poet John Clare wrote about what was being lost, as the whole process came to a head:

> Unbounded freedom rules the wandering scene
> Nor fence of ownership crept in between
> To hide the prospect from the gazing eye
> Its only bondage was the circling sky.[14]

13 Sarah Tarlow, *The Archaeology of Improvement in Britain, 1750–1850* (New York: Cambridge University Press, 2007), p. 50.

14 John Clare, "Enclosures," in Christopher Hampton, ed., *A Radical Reader: The Struggle for Change in England, 1381–1914* (New York: Penguin, 1984), p. 439.

Indeed, dissenters were present all along. Thomas More's *Utopia* (1515–16) contained a bitter indictment of enclosure's toll, at a time when plow land yielded largely to sheep pasture. Wool was the breeding ground of English capitalism.

Poets, commentators, and utopians were not significant factors in the struggle over enclosure. The very long contest has been mostly unreported, in part because most resistance to enclosures was by mainly illiterate peasant-artisans. The standard general histories pay little attention to this movement, and chiefly side with the victors. "Discontent was probably exaggerated," declared Christopher Trent, in *The Changing Face of England.*[15]

Turning to a brief survey of the specifics of resistance, the War of the Roses (1455–85) signaled the end of feudalism, a century ahead of the continent. It also coincided with the opening rounds of a very long war over enclosures: "riotous resistance" occurred in 1469, 1473, 1495, and 1509.[16] 1500 is the conventional date for the inauguration of the "Age of Discovery" (pre-colonialist imperialism); it also marks the age of English enclosures. By the mid-1500s, rancor and violence came to accompany the latter; resistance burst forth. The revolts of 1548–52 (e.g. Kett's rising in Norfolk) centered in eastern and southern England, and were anti-aristocratic as well as anti-enclosure. Kett was an older tanner of Wynmondham, Norfolk, and the rebellion in the summer of 1549 that bears his name not only involved the destruction of ditches and other enclosure barriers, but "shook the foundations of Tudor England," in the words of Barrett Beer.[17]

---

15 Christopher Trent, *The Changing Face of England* (London: Phoenix House Ltd., 1956), p. 111.

16 E.P. Thompson, *Customs in Common* (London: The Merlin Press, 1991), p. 122.

17 Barrett L. Beer, *Rebellion and Riot: Popular Disorder in England during the Reign of Edward VI* (Kent, Ohio: The Kent State University Press, 1982), p. 82f.

Several Tudor rulers, including Henry VIII, were actually opposed to enclosure out of fear of serious unrest by a starving peasantry. In fact, enclosures were being prosecuted in the Star Chamber (England's highest court) as late as 1639, even though government opposition was certainly waning. The very first enclosure by parliamentary act or award had taken place in 1604.

The Midland Revolt of 1607 was a coordinated, armed peasant revolt that spread across Midland counties. A major obstacle to suppressing the uprising lay in the fact that the local militias were unreliable, consisting mainly of plowmen. The revolt was put down and enclosures continued, but so did opposition, e.g., almost continuous riots at Coventry for years to come.[18]

Along with insurrectionary actions, other forms of resistance appeared over the centuries. Organized raids into privatized reserves such as Windsor Forest, Enfield Chase, and Woomer Forest provided game for dispossessed families. Various styles of poaching reflected an underlying refusal to accept the seizure of communal lands. Smugglers were admired and "hardly anyone considered it morally wrong to cheat the Government of taxes."[19] Refusal to pay rents was also a popular phenomenon.

The drive to drain fens or wetlands accompanied enclosure efforts and was met with similar resistance. There were "fierce conflicts" to defend such places throughout the 1600s. At Deeping Fen, for example, about a thousand men destroyed drainage works in 1699.[20]

---

18 John Martin, "The Midlands Revolt of 1607," in Andrew Charlesworth, ed., *An Atlas of Rural Protests in Britain 1548–1900* (New York: Routledge, 1983), p. 36.

19 Christian Hole, *English Home-Life 1500–1800* (London: B.T. Batsford Ltd., 1947), p. 114.

20 Charlesworth, *op. cit.*, pp. 39, 42.

The century may not have been entirely conflictual, and yet it seems that not a single year was without outbreaks of struggle. Thus it was not out of the ordinary that in 1607 Diggers and Levellers made their first challenge, for which their leader was hanged. They are remembered more for their 1649 project of de-enclosing land near Walton-on Thames in Surrey. There Gerrard Winstanley and 50 others held out for over a year in defiance of landlords, Army, and the law. They stood for the abolition of enclosures and of private property itself.

A turning point could already be glimpsed, culminating in the so-called Glorious Revolution of 1688, very far from being either glorious or a revolution. At this time England was in the main a country of commons and of common fields,[21] but 1688 not only greatly strengthened parliamentary government, but also bolstered authority across the board. Whereas earlier enclosures did at least provide some allotments for the poor, or other concessions, this relative leniency was henceforth withdrawn.

Not coincidentally, a move to establish workhouses set in, greatly aided by a parliamentary act in 1722. The East Anglian riots of 1765, among others, opposed the spread of such prisons for the poor. Other privatizing methods included turnpikes and tollgates, which "aroused intense popular opposition."[22] There were riots against turnpikes around Hereford and Worcester in 1727, 1735–36, and 1753, and much destruction of them at times in Bristol, Leeds, Wakefield, and Beeton. Significant anti-enclosure violence continued, including Northamptonshire 1710, Weldon 1724, Forest of Dean 1735, Charnwood Forest 1749, South Fields

---

21 The Hammonds, *op. cit.*, p. 19.

22 George Rudé, *The Crowd in History* (London: Lawrence & Wisehart, 1981), p. 35.

1753, Wiltshire and Norwich 1758, and Oxfordshire 1765, to name just a few.

The growing power of a more activist State turned the tide: parliamentary enclosure acts increased strongly from about 1760. At this time it was no longer deemed necessary for enclosure petitioners to remain quiet about it, and after the 1780s, little more is heard of the case for open-field agriculture and the maintenance of commons.[23] One exception is "Remarks on Enclosure" by "a country farmer" who in 1786 proclaims enclosure a swindle perpetrated against poor folks.[24] The pace of government-enforced enclosure picked up steam and reached its peak in the years between 1793 and 1815.

The extensive—and largely untreated—anti-enclosure riots throughout East Anglia in 1816 show that the struggle was not yet over. The insurgents inscribed "Bread or Blood" on their flag; this is the title of A.J. Peacock's exceptional work on the subject.[25] The serious "Swing" riots of 1830 took place all over agrarian England, especially in the south and east. A principal weapon was arson, aimed at those spearheading enclosure and mechanization. Over 2,000 were imprisoned for the widespread property destruction. 1831 saw open warfare over the fate of the Forest of Dean, with many fences destroyed by hundreds in several Gloucestershire towns.[26] Otmoor, Oxfordshire was the scene of bitter anti-enclosure fighting between 1830 and 1835.

And yet, although powerful, these actions were rear-guard, last-gasp. It is hard to disagree with Michael Reed's assessment that "by the third decade of the nineteenth century the

---

23 Tate, *op. cit.*, p. 86.

24 *Ibid.*, p. 85.

25 A.J. Peacock, *Bread or Blood: A Study of the Agrarian Riots in East Anglia in 1816* (London: Victor Gollancz Ltd., 1965).

26 See Ralph Amstis, *Warren James and the Dean Forest Riots* (Coleford, United Kingdom: Albion House, 1986).

redrawing of the rural landscape embodied in the phrase 'the Parliamentary enclosure movement' was almost complete."[27]

The last major enclosure act was passed in 1845, and the last big riot occurred at Coventry. "Inclosure," as one man told Arthur Young in 1804, during the Napoleonic wars, was "worse than ten wars,"[28] which is why some held out so long and so tenaciously for the old customary economy. Revolt flickered for years (e.g., Berkhamdstead and Epping Forest, 1871), but enclosure had triumphed. As factory slavery developed, resistance to mechanization (e.g., the Luddite risings, 1800–1820) replaced efforts for the land to be returned. Craft work gave way to mass production, along with the compulsory enclosure of all land held in common by the people. The self-supporting cottager who obtained most goods by his own handiwork became a consumer of products in a mass market. The small forms of the old orientation were "turned into factories for bread and meat."[29]

The "primitive" institution of the common was lost. The ardent desire of William Hewitt in 1844, to "chase all commissioners, land-surveyors, petitioning lawyers, and every species of fencer and divider out of their boundaries for ever and ever,"[30] could not be fulfilled. Those enclosed were, of course, non-indigenous; they toiled within already long-domesticated Europe. But the words of Comanche leader Ten Bears seem somehow relevant: "I was born where there are no enclosures and where everything drew free breath. I want to die there. and not within walls."[31]

27 Michael A. Reed, *The Landscape of Britain: From the Beginnings to 1914* (Savage, Maryland: Barnes & Noble Books, 1990), p. 308.

28 Nancy Koehn, "Josiah Wedgwood and the First Industrial Revolution," in Thomas McCraw, *Creating Modern Capitalism* (Cambridge: Harvard University Press, 1997), p. 35.

29 Rowland Prothero, *The Pioneers and Progress of English Farming* (London: Longmans, Green, 1888), p. 161.

30 William Howitt, *The Rural Life of England* (Shannon: Irish University Press, 1971 [1844]), p. 394.

31 Ten Bears, Comanche, Medicine Lodge Creek, 1867. Quoted in Dee Brown, *Bury My Heart at Wounded Knee: An Indian History of the American West* (New York: Sterling Publishing Co., Inc., 2009), p. 274.

MODERNITY TAKES OVER

What were modernity's origins and what has been its trajectory? In this very brief critical survey, let's start with the Renaissance. Ever since Jacob Burkhardt's *The Civilization of the Renaissance in Italy*, the word immediately brings other words to mind: "individual," "self," "personality," usually thought of as modern. Western individualism, and a new age of domination, begin with the Renaissance.

Oswald Spengler used the word "Faustian" to designate a further realm of control. "Renaissance, Rinascita, meant... the new Faustian world-feeling, the new personal experience of the Ego in the Infinite."[1] Of course, he refers here to the so-called Age of Discovery, when emerging European national states reached out to colonize far-flung continents.

"Humanism" is another term that points to the future, with its emphasis on individualism. An individualism that must be seen as corporate, firmly embedded in collective networks of power and wealth. And the "self-confident artistic utopianism of 15th-century Renaissance Florence"[2] existed in the context of a noisy, dirty, violent, typhus- and malaria-ridden city. Along with self-confidence there was much discontent—even despair. As Edgar Wind put it, "the most splendid release of artistic energies was attended by political disintegration."[3]

Merchant bankers dominated the urban politics of the Renaissance, often wielding near-absolute power. In Florence the Medicis amassed huge wealth and authority, but lacked legitimacy. To make interest-bearing loans (usury) was a

---

1   Oswald Spengler, *Decline of the West*, Vol. II (New York: Alfred A. Knopf, 1926), p. 191.

2   Alexander Lee, *The Ugly Renaissance: Sex, Greed, Violence and Depravity in an Age of Beauty* (New York: Doubleday, 2013), p. 42.

3   Edgar Wind, *Art and Anarchy* (New York: Knopf, 1964), p. 6.

mortal sin, Dante's favorite target. And Medici wealth was fairly recently obtained, usually by fraudulent and violent means. It fell to artists to create an artificial aura of legitimacy (e.g., Rubens' Medici cycle of paintings). Patronage of art and architecture succeeded at this task on a grand scale, also in the service of an especially corrupt and violent Papacy (e.g., Alexander VI, Pius II).

The glories of Renaissance culture also papered over a surge of anti-semitism in Florence that "would not be matched in Italy until the rise of fascism."[4] High culture was also deaf to the fact that European expansionism involved consigning entire peoples to non-human status. These atrocities were accompanied by what Alexander Lee termed "the most deadening artistic silence of all time."[5] In fact, mastery of color, perspective, and the like often served the opposite of what we might think of as Renaissance values.

Spengler rightly concluded that "the Renaissance never touched the people."[6] Case in point: the several years' rule over Florence by Dominican friar Girolama Savonarola in the waning years of the 15th century. Simplicity and repentance were the watchwords of his near-revolution. Thousands of youth ran through the streets smashing anything that appeared to be arrogant wealth. It was a virtual theocracy, ISIS-like to some degree, but plebeian in character.[7] Savonarola's social and cultural bonfire did not endure; in 1498 he was hanged, then burned to ashes on the spot.

Half a century later, according to J.B. Singh-Uberoi, "the modern chapter of man and nature as well as of natural science [owed] as much (or more) to the Reformation as it

---

4   Lee, *op. cit.*, p. 308.

5   *Ibid.*, p. 350.

6   Spengler, *op. cit.*, Volume I, p. 233.

7   Thomas Cahill, *Heretics and Heroes* (New York: Doubleday, 2013), p. 79.

did to the Renaissance."[8] Protests against abusive practices of the Catholic Church by Luther, Calvin, Zwingli, and others became a revolt against Papal authority. Protestant denominations were the result: a full break with Catholicism. And the anti-authoritarian spirit of the Reformation was not limited to doctrinal matters. In Germany, home of the Reformation, anger at Church landlords ran high; appeals to the people by Luther and other reformers brought more radical results than these preachers intended.[9] The radical Reformation was exemplified by Thomas Müntzer, who broke with Luther early on, announcing an imminent apocalypse wherein freedom and equality would reign. Müntzer preached dispossession of the nobility, echoing the Taborite millenarians and social revolutionaries of 15th-century Bohemia. The great peasant revolt in southern and central Germany (1525–1526) was the most important event of the Reformation period and one of the biggest mass movements in German history.[10]

But sadly, the Peasants' War is not what was arguably modern about this era. The seeds of modernity are found instead in writings of people like Ulrich Zwingli. He preached the necessity of regular, industrious habits, and warned of "the danger of relaxing the incentive to work."[11] The origin of this modern, now-internalized ethos is the main subject of Max Weber's classic, *The Protestant Ethic and the Spirit of Capitalism*.

The rise of Protestantism relied upon the print culture introduced by Johannes Gutenberg's invention: a printing

8   J.P. Singh-Uberoi, *The European Modernity* (New York: Oxford University Press, 2002), p. 26.

9   Peter Burke, *Popular Culture in Early Modern Europe* (New York: New York University Press, 1978), pp. 260–261.

10   Michael Hughes, *Early Modern Germany, 1477–1806* (Philadelphia: University of Pennsylvania Press, 1992), p. 45.

11   R.H. Tawney, *Religion and the Rise of Capitalism* (New York: Harcourt Brace and Company, 1926), p. 115.

press using movable type. Printed books were available in the early 1500s, accompanied by a striking increase in literacy. For Marshall McLuhan, print was a founding aspect of modernity: "With Gutenberg Europe enters the technological phase of progress, when change itself becomes the archetypal norm of social life."[12]

Typography made possible the first assembly line, the first mass production. Not only did authorial ownership commence, but, according to Roberto Dainotto, "By embedding language in the manufacturing process of mass-produced books, the printing press transformed words and ideas into commodities."[13] Walter Ong observed another key outcome: "Before writing was deeply interiorized by print, people did not feel themselves situated every moment of their lives in abstract, computed time of any sort."[14] A changed sense of time seems related to a growing "passion for exact measurement"[15] in the late Renaissance. The emphasis on precision shows that the domestication process is speeding up and tightening its grip.

The privatization of this medium through silent reading, an enormous change in itself, also altered the balance among our senses. Touch and hearing became much less important. In antiquity and in the Middle Ages, reading was social—reading aloud. Some saw typography as a powerful, alien force. Rabelais and Cervantes declared it "Gargantuan, Fantastic, Suprahuman."[16] Print and literacy led to a marked increase in the social division of labor. Illiterates became

---

12 Marshall McLuhan, *The Gutenberg Galaxy: The Making of Typographic Man* (Toronto: University of Toronto Press, 1962), p. 155.

13 Roberto M. Dainotto, *Europe (in theory)* (Durham, NC: Duke University Press, 2007), p. 35.

14 Walter J. Ong, *Orality and Literacy* (London: Methuen, 1982), p. 96.

15 McLuhan, *op. cit.*, p. 166.

16 *Ibid.*, p. 194.

subordinates, subject to the greater effective power of specialists, and witness to a steady dissolution of community.[17] Community became less important than one's place in the division of labor hierarchy.

A cognate of *public* is *publish*. We come now to a foundation of mass society: mass media as a means of social control. Print greased the wheels for national uniformity and state centrism; yet at the same time it facilitated individual expression and opposition to the dominant order.

Humanism is the watchword of Renaissance thinking. *Humanitas* is its Latin reference, opposed to *immanis*, or savage. Humanism's proponents stressed individualism, but that covered a multitude of sins. An individualist spirit of inquiry and adventure helped fuel overseas invasions and territorial expansion. Humanists were often silent about the deeds of colonialist explorers, but occasionally there was a direct connection. Amerigo Vespucci, for instance, was an explorer and a humanist writer. He had also worked for the Medici bank in Florence.

Many humanists sanctioned the subjugation of women. Renaissance power was inherently masculine. During the Renaissance period, women lost status compared to their medieval sisters. although women of the middle and poorer classes retained more self-determination than those of higher rank.[18] Between 1480 and 1700 (the heyday of humanism), large numbers of women were condemned and executed as witches.

Renaissance humanists were filled with zeal to rekindle the Crusades and wipe out Muslims. The early and

---

17 See Agnes Heller, *Renaissance Man* (Boston: Routledge & Kegan Paul, 1978), e.g., pp. 206, 291.

18 William Caferro, *Contesting the Renaissance* (Malden, MA: Wiley-Blackwell, 2011), pp. 15, 68, 76, 86.

much-cited humanist Petrarch was especially venomous against Islamic infidels.[19]

As always, intellectuals were called upon to legitimate the dominant order, and humanists performed this service. Michel de Montaigne, a 16th-century magistrate and essayist, is principally noted for his project to question everything—a very modern idea. He is seen as the first fully humanist writer, the first to express a coherent version of the doctrine. He denied that commoners and women could engage in the search for self-knowledge.[20]

The earthy François Rabelais, Montaigne's contemporary, was the rare antinomian figure of the period. His utopian Abbey of Thélème was a place of pleasure and freedom, not of sanctioned individualism.

As 1600 approached, humanism's legitimacy was challenged. "Late humanism was beset with a crisis of confidence," in Katherine Eggert's words.[21] Something seemed to be missing. Something human was being lost. But what would take its place?

In 1582 time was brought up to date with the introduction of the Gregorian calendar, which reigned over a time of "a general malaise"[22] and active disaffection in Europe. The late 16th century was marked by serious peasant revolts. France and the Netherlands experienced urban disorder, not forgetting the great 1585 rising in Naples. In 1600 Giordano Bruno was burned at the stake in Rome's Campo Fiori for defending Copernicus and espousing dangerous ideas of atomic theory and an infinite universe.

---

19 Lee, *op. cit.*, pp. 318–319.

20 Caferro, *op. cit.*, p. 33.

21 Katherine Eggert, *Disknowledge* (Philadelphia: University of Pennsylvania Press, 2015), p. 8.

22 Alexander Cowan, *Urban Europe 1500–1700* (New York: Oxford University Press, 1998), p. 189.

Seventeenth-century thinkers dethroned scholastic Aristotelianism and, indeed, theology itself. Not only Church orthodoxy, but animism and magic that had survived into the Renaissance were rejected. The mental universe was still animate rather than mechanical, though, despite the concept of conquest of nature whose roots lay in the Renaissance.

The scientific revolution of the 1600s was a decisive break with the past, a thorough re-evaluation of what had come before. Francis Bacon (1561–1626) has come to represent the shift. Inaugurating methods of induction and experimentation, his project was to restore the dominion over creation that had been lost with the expulsion of Adam and Eve from the Garden of Eden. Bacon saluted America's first colonizers, their work in a "Newfound Land of inventions and sciences unknown."[23]

But Bacon did not achieve a full break with Church scholasticism (of Thomas Aquinas and others). That task fell to René Descartes, and Michel Serres' words are worth noting: "Mastery and possession: these are the master words launched by Descartes at the dawn of the scientific and technological age, when our Western reason went off to conquer the universe. We dominate and appropriate it: such is the shared philosophy underlying industrial enterprise as well as so-called disinterested science, which are indistinguishable in this respect."[24]

A self-proclaimed original, Descartes was an arch-rationalist who refused to trust his own senses. His dis-embodied approach sought to derive sensory information from mathematics instead of the other way around, and virtually equated math and natural science. Having created analytic

23 Loren Eiseley, *Francis Bacon and the Modern Dilemma* (Lincoln, NE: The University of Nebraska Press, 1962), p. 59.

24 Michel Serres, *The Natural Contract* (Ann Arbor, MI: University of Michigan Press, 1995), p. 32.

geometry, he wanted to mathematize thought. Descartes' famous formulation of mind-body dualism is consonant with his view of reality as immutable and inflexible mechanical order. It should come as no surprise that he saw humans, among other living beings, as fundamentally machines.

The Cartesian project did much to initiate modern thought and at base, still obtains. Now we witness the Artificial Intelligence technicians striving for Artificial Consciousness, pursuing a machine model. And contemporary philosophy seems to take seriously hyper-estranged Alain Badiou's mathematics-equals-ontology concept. Descartes subverted humanism, and gravely worsened the un-health of the West.

Among Descartes' contemporaries was Gottfried Leibniz, whose new system of "pre-established harmony" offered a mechanistic explanation of Creation—a further move in the onslaught on scholasticism. John Locke, founder of the modern liberal, individualist tradition, rejected Descartes' dualism as too God-oriented. Locke attacked political absolutism for the non-productivity of the land-owning aristocracy. He argued for a more modern form of exploitation, the enclosure of communal land into privately owned property. The 17th-century backdrop to these published ideas was burgeoning occupation and enslavement on other continents by European profiteers. Thomas Hobbes was party to this through his involvement with the Virginia Company. He condemned life in the state of nature as "nasty, brutish, and short," and termed indigenous people "savages," providing ideological justification for conquest and slavery.

Price hikes and tax increases provoked resistance, such as the 1630 rising in Dijon and revolts in Aix-en-Provence between 1631 and 1638. Silk workers in Amiens attacked their masters' establishments in 1637. Bayeux tanners rose up briefly in 1639, and sailors' wives went on the offensive in Montpelier in 1645, to cite a few insurrectionary incidents in 17th-century France.[25] During this period the Thirty Years' War (1618–1648), Europe's last mainly religious war, ravaged a third of the continent, with millions of casualties.

1648 was a year of revolts, particularly in the context of the English Civil War. Levellers, Ranters, Diggers, and others espoused radical, anti-authority, anti-enclosures orientations. But Oliver Cromwell's Protectorate prevailed over the resistance, establishing mercantile capitalism as the core of the economy.

By this time, and commencing in earnest around 1600, division of labor was transforming the ground of social existence.[26] New production techniques ushered in proto-industrialization, especially in rural areas.[27] "Proto-industries arose in almost every part of Europe in the two or three centuries before industrialization."[28]

The ideas of Bacon, Descartes, Leibniz, and other mathematical and scientific thinkers interwove with and supported technological innovation during the 17th century. As Margaret Jacob notes, "The road from the Scientific Revolution to the Industrial Revolution...is more straightforward than we may have imagined."[29]

What we call the Enlightenment of the 1700s owed much to the canon of 17th-century empirical philosophy and natural science. Denis Diderot's iconic *Encyclopédie* was based on his

25 Cowan, *op. cit.*, pp. 44, 176–179.

26 John Zerzan, *Future Primitive* (Brooklyn, NY: Autonomedia, 1994), p. 147.

27 Peter Kriedte, Hans Medick, Jurgen Schlumbohm, eds., *Industrialization Before Industrialization* (New York: Cambridge University Press, 1981).

28 Sheilagh C. Ogilvie and Markus Cerman, "Proto-industrialization, economic development and social change in early modern Europe," in Ogilvie and Cerman, eds., *European Proto-Industrialization* (Cambridge: Cambridge University Press, 1996), p. 227.

29 Margaret C. Jacob, *The Cultural Meaning of the Scientific Revolution* (New York: McGraw-Hill, 1993), p. 7.

"tree of knowledge," derived from Bacon's 17th-century ideas.[30] Although initially an English phenomenon, Enlightenment is best known for its flowering in Paris, between the death of Louis XIV in 1715 and the onset of the French Revolution in 1789. Its most important figures were Voltaire, Montesquieu, and Rousseau.

During this period, protests and riots continued to flare (e.g., Geneva experienced risings in 1717, 1738, 1768, and 1782). Newspapers and commercialized leisure became part of everyday life. In the 1750s and 1760s the modern chronological timeline was first introduced,[31] and modern education forms (such as measurable results via written examinations) became common.

Enlightenment voices decried superstition and tyranny. Christianity came under fire, most forcefully by the programmatic disbelief of Diderot and David Hume, among others. The Church retreated, dissolving the militant Jesuit order (it would not be re-established until 1814). The new outlook overturned the Renaissance belief that what came first was best, replacing it with faith in progress and the future. A favorite target of Enlightenment's materialist orientation was animism; the once-prevailing conception of a living spirit in nature was denounced as superstition.

The supposed anti-tyranny credo bears a closer examination. Voltaire and other leading Enlightenment lights were friendly with Frederick the Great, despite his despotism and support of feudalism.[32] Frederick's proclamation

---

30 Jeffrey S. Ravel and Linda Zionkowski, *Studies in Enlightenment Culture*, vol. 36 (Baltimore: The Johns Hopkins University Press, 2007), p. 55.

31 Robert Darnton, *The Great Cat Massacre and Other Episodes in French Cultural History* (New York: Basic Books, 1984), p. 194.

32 Alexander Rüstow, *Freedom and Domination: A Historical Critique of Civilization* (Princeton, NJ: Princeton University Press, 1980), p. 330.

of the Enlightenment as Prussia's official ideology[33] seems like a strange fit.

Enlightenment reason certainly did some demythologizing, but it also installed new myths along with its claims and promises. One such myth held that history, in Couze Venn's words, as "the universal and rational project of the becoming of humanity as a whole"[34]—a myth with grave implications and consequences for indigenous people. There is an evident connection between imperialist expansion as a system of power, and the diffusion of Enlightenment thought as a global pattern of culture.

A forebear of dissent from the vision of universalizing Progress was Rabelais. He declared his "enduring affinity with the alien spirits, of whom there are always some in every society, who at any sacrifice resist, or rather, quietly elude, all pressure towards conformity, towards standardization and mechanization of thought."[35] In his utopian Abbey of Thélème, there are no clocks; a swimming pool and other non-monastic features are inspired by the abbey's all-encompassing watchword, Do What Thou Wilt. Eighteenth-century philosopher and novelist Jean-Jacques Rousseau took a dim view of civilization and proclaimed the natural goodness of humankind. He refused abstract geometry and its method, preferring the promenade as a way of visiting "idle and lazy" nature, as Michel Serres put it.[36]

---

33 T.C.W. Blanning, *The Culture of Power and the Power of Culture* (New York: Oxford University Press, 2002), p. 201.

34 Couze Venn, "Altered States: Post-Enlightenment Cosmopolitanism and Transmodern Socialities," in *Theory, Culture & Society* 19:1 (2002), p. 65.

35 Albert Jan Nock and C.R. Wilson, *Francis Rabelais* (New York: Harper & Brothers Publishers, 1929), p. 358.

36 Pierre Saint-Armand, *The Pursuit of Idleness: An Idle Interpretation of the Enlightenment* (Princeton, NJ: Princeton University Press, 2011), p. 14. Similar cases Chardin's distraction, Rameau's vagrancy, Joubert's laziness, p. 16.

Central to Enlightenment thought and probably the most important modern philosopher, Immanuel Kant did much to shape how people understand reality even today. He also revealed something of the less than liberatory side of Enlightenment. Silke-Maria Weineck placed his thinking "on the side of certifiable calculations, of the exchange of goods, of sound economics."[37] Similarly, Heinrich Heine referred to Kant's "petty-bourgeois values."[38] Theodor Adorno took this further, observing that "All the concepts... [Kant's] Critique of Practical Reason proposes, in honor of freedom—...law, constraint, respect, duty—all of these are repressive."[39] He found that "Kant's moral philosophy...will not let him visualize the concept of freedom otherwise than repression." And "reason itself is to Kant nothing but the lawmaking power.... He glories in an unmitigated urge to punish."[40] Montesquieu is closely aligned with Kant in this: "Law, generally speaking, is human reason."[41]

The empire of Reason also liquidates difference, in the direction of the "total, perfect political unification of human species," in Jacques Derrida's words[42]—a cold, universalizing agenda. The German poet Novalis found a conformist spirit of disenchantment in the "harsh, chilly light of the Enlightenment."[43]

---

37 Silke-Maria Weineck, *The Abyss Above* (Albany, NY: State University of New York Press, 2002), p. 10.

38 Anthony Pagden, *The Enlightenment: Why It Still Matters* (New York: Random House, 2013), p. 387.

39 Theodor Adorno, *Negative Dialectics* (New York: Continuum, 2007), pp. 230, 232.

40 *Ibid.*, pp. 255, 256, 260.

41 In John W. Yolton, et al., eds., *The Blackwell Companion to the Enlightenment* (Cambridge, MA: Blackwell, 1991), p. 258.

42 Quoted in Venn, *op. cit.*, p. 74.

43 Ulrich Imhof, *The Enlightenment* (Cambridge, MA: Blackwell, 1994), p. 270.

Its supposed higher form of rationality provided cover for Europe's "civilizing mission" and for Western hegemony. Without the new imperialism, as Paulos Gregorios saw it, the Enlightenment "could hardly have taken place."[44] Ideas and actions deeply influence each other.

In France the Enlightenment emerged after the reign of Louis XIV, when it "began to set the tone in polite society."[45] Enlightenment *philosophes* felt confidence, at least in part, because of their close relationships with bourgeois notables. Not only Frederick the Great, but other ministers and sovereigns looked to them for guidance and legitimation. The patronage of absolutist princes created influential positions for them; as J.B. Bury reminds us, "They never challenged the principle of a despotic government, they only contended that the despotism be enlightened."[46] Before the Revolution that began in 1789, Enlightenment standard-bearers were "part of the new ruling elite."[47]

It is also true that the modern understanding of citizenship is a creation of the Enlightenment. But as Voltaire said, "Better not teach peasants how to read; someone had to plow the fields."[48] Yet Voltaire also passionately denounced slavery, as did Condorcet and Raynal. There were also protests against the oppression of colonial peoples, though not against the practice of colonization itself.

Summing up their mid-20th-century critique of the Enlightment, Max Horkheimer and Theodor Adorno declared that the "fully enlightened earth radiates disaster

---

44 Paulos Mer Gregorios, *A Light Too Bright: The Enlightenment Today* (Albany, NY: State University of New York Press, 1992), p. 63.

45 Darnton, *op. cit.*, p. 208.

46 J.B. Bury, *The Idea of Progress* (New York: Dover Publications, 1955), p. 176.

47 Robert Darnton: *The Case for Enlightenment: George Washington's False Teeth* (New York: W.W. Norton, 2003), p. 5.

48 R.G. Saisselin, "Philosophes," in Yolton et al., *op. cit.*, p. 397.

triumphant."[49] Modern exploitation of nature and modern, atomized mass society commence with this epoch.

Enlightenment thought was an ideological bridge between a pre-industrial, aristocratic culture and an industrialized, consumerist society. Some large-scale production facilities existed in mid-18th-century Europe: examples include van Robais' textile factory at Abbeville, the Lombe brothers' silk mill at Derby, and the iron industry initiated by Peter the Great in the Ural Mountains.[50] Some Enlightenment proponents were directly involved in these enterprises. Diderot studied the mechanical order of production; Vaucanson designed efficient silk mills. Early manufacturers breathed the air of the dominant liberal, humanistic creed of the Enlightenment. Its spirit of classification and analysis was a practical aid to industry. Enlightenment materialism fostered "mastery by technological and commercial means over the material world."[51]

The principle of individual autonomy, even with the necessary qualifiers, gained acceptance during the Renaissance and the Protestant Reformation. But as Bruno Latour argues, "modern" only applies to societies in which artisanal, personal kinds of making are superseded by broad-scale, impersonal modes.[52] Modernity is an Enlightenment word, and Latour's watershed distinction can be found in that era. The Enlightenment was the first take-off point of the non-conscious praxis of amoral technicism.

---

49   Theodor Adorno and Max Horkheimer, *Dialectic of Enlightenment* (New York: Verso, 1997), p. 3.

50   Norman Hampson, *A Cultural History of the Enlightenment* (New York: Pantheon Books, 1968), p. 46.

51   Aram Vartanian, "Diderot and the Philosophes," in Yoltan et al., *op. cit.*, p. 316.

52   Julian Yates, *Error, Misuse, Failure* (Minneapolis: University of Minnesota Press, 2003), p. 5.

Major claims and promises were made. There would be an end to religious intolerance, and a Brave New World ushered in by science and technology. Given the evident failure of these promises, it is little wonder that there is now "a global backlash...against the Enlightenment itself," as John McCumber has put it.[53] I concur with Onora O'Neill's assessment: "A world of isolated and alienated individuals who find to their horror that nihilism, terror, domination, and the destruction of the natural world are the true off-spring of the Enlightenment."[54]

We are still in the Enlightenment era, and its "light" is spreading everywhere. The fully enlightened world, the fully civilized world, is indeed disaster. The prospect of modernity without end faces each and all of us.

---

53 John McCumber, *On Philosophy: Notes from a Crisis* (Stanford, CA: Stanford University Press, 2012), p. 10.

54 Onora O'Neill, "Enlightenment as Autonomy: Kant's Vindication of Reason," in Peter Hulme and L.G. Jordanova, *The Enlightenment and its Shadows* (New York: Routledge, 1990), p. 189.

DIVIDED LIFE

In Greek myths we find figures like Daedalus, Medea, Jason, and Hephaestus, who had miraculous powers to project utopian technological possibilities. Adrienne Mayor's *Gods and Robots: Myths, Machines, and Ancient Dreams of Technology*[1] is a very good resource here. Hard to miss the parallels with the present, as technology is touted as what can save us, heal us, secure our future and that of the planet.

The seeds of such mythic fantasy figures were planted well before Greek civilization. It is important to see the foundation of such thinking as a function, at base, of the most primary of social institutions: division of labor.

With division of labor or specialization, a reconstitution of human existence is set in motion. Its development, very slow at first, enabled the quantum leap to domestication. Specialists, spiritual and otherwise, introduced differentials of authority, leading to the full-blown control ethos of domestication.

Collective, cooperative relations become distorted by division of labor, and community begins its long decline. Specialization requires exchange, not always reciprocal. The ethos of sharing, hallmark of pre-civilization band society, faces the challenge of a different direction for society. Tools and the autonomy they represent are ultimately overtaken by systems of technology, which require management.

Stepping further back for a moment, it is language that allows for effective division of labor, for the decisions and coordination it involves. It is possible, maybe even likely, that gendering was making its appearance along with speech. Date(s) unknown, but probably somewhat recently in terms of the long span of the *Homo* species. That is, for example,

---

1   Adrienne Mayor, *Gods and Robots: Myths, Machines, and Ancient Dreams of Technology* (Princeton, NJ: Princeton University Press, 2018).

the *idea* of man as distinct from that of simply male. Here, arguably, is the most fundamental division of labor, the sexual one, the cultural roles that are assigned by gender.[2]

It is with the material reproduction of society that we find a basic and graphic "case study" of what division of labor produces. The increasing division of labor is the hallmark of industrial social existence. It divides and deadens the self—and society. In this important respect, technology is not a tool, but a world and a worldview. It is ever more determining and pervasive; and driven, as ever, by division of labor.

The young Georg Lukacs, in such works as *Soul and Form* and *Theory of the Novel*, pointed out the destruction of individual expression represented by the division of labor, and a collapse of community into atomized and isolated subjects. His contrary model was the artistic German craftsman tradition.[3] The degradation of work occupied the likes of William Morris, John Ruskin, and William Wordsworth before Lukacs, and of course many others today, including poet and novelist Marge Piercy, as well as numerous sociologists—not to mention the resistance of workers themselves in every era.

Labor has been progressively denuded of meaning and has been disastrous to human dignity. Talcott Parsons, at least in his early writings, was much impressed with the continued opposition to high levels of differentiation and specialization in modern society.

But division of labor itself, as we have seen, predates the past century or two considerably. In fact, it is at least as old as civilization. Early mining, in both Mesopotamia and Egypt,

---

2   Kate Young, Carol Walkowitz and Roslyn McCullough, *Of Marriage and the Market* (London: CSE Books, 1981); Kate Millett, *Sexual Politics* (London: Abacus/Sphere, 1972).

3   Axel Honnuth, *The Fragmented World of the Social* (Albany, NY: State University of New York Press, 1995), pp. 54, 56.

was conducted via "a minute division of labor,"[4] in Lewis Mumford's words. Very similar to that of the first armies in those first civilizations in the West. And not confined to either of those realms: when Herodotus visited Egypt in the fifth century B.C., he witnessed specialism comparable to that of today. For example, "some physicians are for the eyes, others for the head, others for the teeth, others for the belly, and others for internal disorders."[5] Neither was the assembly-line-type division of labor a 20th-century development. Dante and Marco Polo in 1260 in Venice saw warships built and fitted out as they were moved along a canal with workers stationed on either side.[6]

Franz Borkenau's *The Transition from the Feudal to the Bourgeois World View* (1934)[7] detected a decisive factor behind the major shift in consciousness in the West in the 17th century. He pointed to a great extension of division of labor from about 1600 that introduced the notion of abstract work. It is easy to see this as the wind in the sails of Descartes in that century. The momentous Cartesian mind-body split parallels the disembodiment taking place vis-à-vis work, in everyday life. This development and its philosophical counterpart enabled the arrival of industrialism in the 18th century. They are its foundation.

Marx saw in the herding of people into the new factories the creation of a unified, disciplined proletariat. Of course, it was actually a Great Leap Forward in domestication, pacification, control. Early on, Marx vehemently opposed division of labor, the essence of the factory ethos. He dropped that

---

4   Lewis Mumford, *The Myth of the Machine: Technics and Human Development* (New York: Harcourt, Brace & World, 1967), p. 240.

5   *Ibid.*, p. 193.

6   Richard Donkin, *Blood, Sweat and Tears: The Evolution of Work* (New York: W.W. Norton, 2002), p. 144.

7   English translation. Not published in English.

upon embracing the Industrial Revolution. In *Capital* he admitted that "Some crippling of body and mind is inseparable even from division of labor in society as a whole."[8] Marx went so far as to quote Adam Smith, godfather of the modern division of labor, in an unflinching concession as to its cost: "To subdivide a man is to execute him, if he deserves the sentence, to assassinate him if he does not.... The subdivision of labor is the assassination of a people."[9]

For Marx, this pathology must stand—and be extended. This is the pivotal move against freedom and health. Now what is decisive is who is in charge. As Marxist scholar Moishe Postone put it, regarding the alienation that is division of labor, "its overcoming entails its appropriation by people—rather than its simple abolition."[10] Fellow Marxist Andre Gorz endorses the distinction: "The capitalist division of labor is the source of all alienation."[11] With its "appropriation by people," that is under a socialist regime, lies the answer. No alienation there! Gorz goes on to advocate "new forms of co-operative management" rather than "an unconsciously guided rebellion against the division of labor."[12]

Such "unconsciously guided rebellions" can be a challenge to industry and government. Michael Seidman's *Workers Against Work: Labor in Paris and Barcelona during the Popular Fronts*[13] explores this phenomenon in mid-1930s France and Spain. Seidman found that resistance to industrial discipline

---

8 Karl Marx, *Capital: An Abridged Edition* (New York: Oxford University Press, 2008), p. 224.

9 *Ibid.*, p. 225.

10 Moishe Postone, *Time, Labor and Social Domination* (New York: Cambridge University Press, 1993), p. 165.

11 Andre Gorz, editor, *The Division of Labor* (London: Humanities Press, 1976), p. vii.

12 *Ibid.*, p. 143.

13 Michael Seidman, *Workers Against Work: Labor in Paris and Barcelona during the Popular Fronts* (Berkeley: University of California Press, 1991).

continued unabated under leftist management, despite its new left-wing appropriation. His book's title speaks volumes.

Civilization's advance has always meant more work, and its increasing control dynamic is centrally achieved by reducing work skills. There may be increased political rights and freedoms, but deskilling renders the individual more subservient. Even the semblance of autonomy or self-sufficiency is taken away.

Gone is craft. There's Computer Aided Design (CAD), Semi-Autonomous Mason (SAM, robotics), teachers consigned to drilling students for standardized tests. Where are individuality, creativity, the making of quality things? Handwriting (cursive) is abandoned in favor of a life of typing at the computer. The preponderance of jobs exists in the service and information sectors, requiring minimal skills.

But what about all the labor-saving devices we are blessed with? New technology creates new "needs." With the vacuum cleaner, for instance, the need to have a cleaner house. The general aversion to dirt, by the way, has ushered in less robust, less disease-resistant moderns. "Labor-saving" is a grave misuse. There is more work along the line, not less. Both before and after the production of such a device, the total work involved—not to mention the assault on the natural world—is greater.

To return to tools, to abandon systems of technology, would be romantic and reactionary, we are told. "Going back to the spinning wheel is obviously absurd," counsels Stephen Marglin.[14] After all, one cannot question the validity of the mass production to mass society/culture trajectory. The more reasonable approach is "the possibility of designing computer artifacts in such a way that using them resembles

---

14 Stephen A. Marglin, "What Do Bosses Do?" in *Review of Radical Political Economics*, 6:60 (July 1974), p. 61.

craftsmanship," according to Pelle Ehn.[15] That their *use* may resemble craftsmanship? It is their division of labor production that has defeated craft and determines society.

There is "no exiting from any currently obtaining technology," avers Peter Hershock, evidently avoiding the question as to the nature of its foundation.[16] Division of labor's basic role regarding social existence is rarely discussed. For example, Peter Wagner's *A Sociology of Modernity: Liberty and Discipline* (1994)[17] does not accord it a mention.

The techno-world and its specializations in every sphere may present itself as if somehow a global reality that came about as a kind of mutation. But the motor of this reality has been there all along, and keeps revealing itself. Where efficiency, for instance, is the highest value of complexifying society, one asks what is the value of what is actually gained and what is sacrificed. What is memorable?

VICE (January 7, 2020)[18] reports that farmers are buying 40-year-old tractors that don't have specialized software and thus can be repaired. There is some urge to reskill, from small-scale craft distilling to handicrafts, and the acquiring of primitive or earth skills. Small signs, but maybe a kernel of growing opposition to our divided selves in a divided life-world.

---

15 Pelle Ehm, *Work-Oriented Design of Computer Artifacts* (Stockholm: Arbetslivscentrum, 1988), p. 372.

16 Peter D. Hershock, "Turning Away from Technotopia," in Peter D. Hershock, Mariette Stepaniants, and Roger T. Ames, *Technology and Cultural Values* (Honolulu: University of Hawaii Press, 2000), p. 593.

17 Peter Wagner, *A Sociology of Modernity: Liberty and Discipline* (New York: Routledge, 1994).

18 Matthew Gault, "Farmers Are Buying 40-year-old Tractors Because They're Actually Repairable," *VICE*, January 7, 2020.

TWILIGHT OF THE EVENING LANDS: THE CASE OF OSWALD SPENGLER

Oswald Spengler (1880–1936) was a reactionary, might-makes-right German nationalist. He also came up with a remarkable work of metahistory or theory of civilization. As Karl Marx said somewhere, referring to novelists, one often learns more about society from the reactionary writers than from progressives.

Des Untergang Des Abendlandes is known in English, rather less poetically, as *The Decline of the West*. Its two volumes were mainly written just before and during World War I. Something of a bombshell in the 1920s, as Germans and other combatants reeled in shock from the carnage of the war, the book fell out of favor thereafter. Frank Ankersmit wrote in 2005 that the work was "now sadly underestimated."[1] Earlier, Theodor Adorno had provided a more specific comment on one part of Spengler's contribution: "...the course of world history vindicated his immediate prognoses to an extent that would astonish if they were still remembered."[2] Alfred Stern points out a prognosis of some pertinence today: Spengler's conclusion that cultures/civilizations (he used the terms fairly synonymously) last about a thousand years, and that this ("Faustian") one began with the Crusades. The first Crusade dates from 1095—forecasting the end of civilization in 2095(!).[3]

Spengler looked at eight civilizations, arguing that each has its life cycle, its spring, summer, fall, and winter. He

---

1   Frank Ankersmit, *Sublime Historical Experience* (Stanford, CA: Stanford University Press, 2005), p. 309.

2   Theodor Adorno, *Prisms* (London: Neville Spearman, 1967), p. 54. Donald R. Kelley, *Frontiers of History* (New Haven, CT: Yale University Press, 2006), p. 206, provides a contrary overall assessment, calling Spengler "sensational and largely discredited."

3   Alfred Stern, *Philosophy of History and the Problem of Values* (The Hague: Morton, 1962), p. 127.

pegged the onset of decline/winter in the West as early as 1800. Every civilization has come and gone in this way, obeying a kind of cosmic destiny. The fatal fulfillment of life cycles is inescapable, corresponding to the general pattern of living organisms. One may well wonder whether the fatalities are more like suicides than natural deaths. But the appeal of his overall concept is manifold. A colonized person, for instance, might well be drawn to a book entitled *The Decline of the West*. Today, a looming sense of crisis, decadence, and tragedy gives the work a general, even compelling appeal.

He was very influenced by Friedrich Nietzsche, who preceded him by almost half a century. Spengler was an obscure high school teacher who, like Nietzsche, never held a university position or had academic standing. Both men died at 56.

He held that war was good for the vitality of a culture, and that hierarchy was the natural order of things. His minor book *Prussianism and Socialism* (1920) advocated a form of state socialism based on obedience and hard work. It resembled the Nazis' later National Socialism structure to some degree, but Spengler hated the Nazis for their anti-Semitic rubbish. His opposition to the Hitler regime became quite well known. Had he not died of a sudden heart attack in 1936, he would have most likely ended up in a concentration camp.[4]

A German jingoist, Spengler, like Hegel, saw in the German nation the chosen summit of the World Spirit. But unlike Hegel's abstract works, *The Decline of the West* is breathtakingly rich in historical particularity. Others wrote about the cycles of civilizations: Giambattista Vico in the early 1700s, and even Florus of Rome, around 125 C.E. Spengler greatly surpassed their basic approach, with his breadth of vision and richness

---

4   Klaus P. Fischer, *History and Prophecy* (New York: Peter Lang, 1989), pp. 74, 79.

of references. Nietzsche's example encouraged Spengler to take an outsider's position, but Goethe was far more influential. Spengler more or less transposed Goethe's principle of plant morphology to human history and culture.

For Spengler, every civilization is unique and self-sufficient, so there can be no "timeless" art. In fact, realities of time and space are also defined or experienced differently in various cultures. But the striking, even brilliant contribution Spengler makes is his comparative overview of civilizations. That is, each stage of a culture corresponds with its counterpart in other cultures or civilizations. They are homologous or equivalent, expressions of the same life-cycle period. Their cultural significance is the same; Alexander the Great and Napoleon are contemporaries in this sense. Greek sculpture and the works of Haydn, the pyramids of Fourth Dynasty Egypt and Gothic cathedrals, Indian Buddhism and Roman stoicism, Pythagoras and Descartes, and so on. A staggering breadth of scholarship is prerequisite for such an overview.

Our present, global, Westernized civilization of technology and capital he called Faustian, another borrowing from Goethe. Its prime symbol is limitless space; geographical expansion of power and influence is a cardinal aspect of every late culture. Music, which transcends temporal and spatial limits, embodies the Faustian spirit reaching toward infinity. Spengler noted that the last stage of every civilization has its main achievements in technology, not in, say, the arts or philosophy—Rome, for example. He also saw the negative contemporary impacts of media and increasing urbanization.

It is impossible to miss the fact that the 20th-century German critique of modernity was exclusively a rightist phenomenon. Along with Spengler, others come to mind: Martin Heidegger, Ludwig Klages, Ernst Junger, Carl Schmitt, Hans Freyer. Klages' 1913 lecture "Man and Earth" was anti-civilization, as well as ecological. Spengler wrote that mankind "is an element of all-living nature that rises in rebellion against nature. He will pay for his defiance with his life."[5] Walter Benjamin made a great effort to bring such thinking to the Left, but was thwarted by the progressivist core of leftism.

The decay and failure of all civilizations heretofore can easily be discerned, but what is it that sets civilization itself into motion? Arnold Toynbee pointed out that Spengler had nothing to say about the genesis of cultures, a defect, he felt, "unworthy of Spengler's brilliant genius."[6] For Spengler, the question of civilization's origins had to do with "soul" rather than anything so pedestrian as the development of such basic social institutions as division of labor and domestication. Joseph Tainter, Jared Diamond, and others are useful in filling in that blank.

Late in life, Spengler realized that pre-civilized humanity was a very important area he had overlooked.[7] He had long admired and respected what he saw as the eternal primitive of the species, its true, natural essence. He regarded every symbolic artifact as merely "a sublimation of the eternal primitive."[8] "Domestic animals are civilized animals," he averred. "Man makes them to resemble himself."[9] His domesticated self, it might be added.

But death intervened in this late-inning effort. We will never know how Spengler's political orientation might have changed if he had had more time.

---

5   From *Urfragen: Fragmente aus dem Nachlass*, quoted in John Farrenkopf, *Prophet of Decline* (Baton Rouge, LA: Louisiana State University Press, 2000), p. 222.

6   William H. McNeill, *Arnold J. Toynbee: A Life* (Oxford: Oxford University Press, 1989), p. 1.

7   Charles M. Fair, *The Dying Self* (Middletown, CT: Wesleyan University Press, 1969), p. 43.

8   Fischer, *op. cit.*, p. 121.

9   Oswald Spengler, *Aphorisms* (Chicago: Henry Regnery, 1967), p. 118.

DECADENCE AND THE MACHINE

Fin-de-siècle Europe from 1880 to 1900, and especially the 1890s is known as a period of cultural and social Decadence. This term is somewhat elusive, though there are at least a few parallels to our own time. The philosopher C.E.M. Joad went so far as to conclude, "There is not, I think, any word whose meaning is vaguer, and more difficult to define."[1]

In the arts as in bohemia in general, one thinks of Baudelairian dandyism, irreverent wit, a cultivated languor. Oscar Wilde comes to mind. As R.K.R. Thornton put it, "some young men in various countries...call themselves Decadents, with all the thrill of unsatisfied virtue masquerading as uncomprehended vice."[2] There is little doubt that Decadence made 20th-century modernism possible by breaking long-standing strictures and conventions.

More basically, Decadence was a darkening disillusionment that pervaded thought, imagination, and material life, and which was inseparable from the triumph of industrialism. The characteristic sense of general decay flowed from being lost in the darkness of a completely ascendant Industrial Revolution. In the words of Nietzsche, "Many chains have been laid upon man.... He suffers from having worn his chains for so long."[3]

An unsettled time of doubt, but more than that—an epoch when defeat was deeply felt. "This sense of unrest, of disease, penetrates down even into the deepest regions of man's

---

1   C.E.M. Joad, *Decadence: A Philosophical Inquiry* (London: Faber and Faber Ltd., 1948), p. 55.

2   Ian Fletcher, ed., *Decadence and the 1890s* (Teaneck, NJ: Holmes and Meier, 1978), R.K.R. Thornton, "'Decadence' in Later Nineteenth-Century England," p. 17.

3   Friedrich Nietzsche, *Human, All Too Human* (New York: Cambridge University Press, 1986), p. 393.

being," according to Edward Carpenter.[4] When the world presents itself as a mechanism of impersonal forces beyond human control, a done deal, with the full connivance of both Left and Right, Decadence is unavoidable. Ennui reigns; only technology is dynamic. Everything healthy is in decline and "the decadent mentality resigns itself to awaiting it passively, with anguished fatality and inert anxiety."[5] Sound familiar? The ethos of failure was palpable. The chief power of the era was that of industrialism, hands down.

In the first half of the century E.T.A. Hoffmann, Mary Shelley, and Edgar Allan Poe shuddered at automatons and other mechanical figures as if they saw in them the future reality of industrializing humanity. There was of course a persistent Romantic critique of mechanistic Progress. From about 1830 when the impact of the factory was really registering, various oppositional voices were heard, including Zola, Balzac, and Flaubert in France; Heine, Hesse, and Thomas Mann in Germany; Carlyle, Dickens, Ruskin, Morris, and Carpenter in England; and Tolstoy in Russia, to name a few.

But Decadence was not an extension of Romanticism but a reaction to it. In the absence of significant anti-industrial movements and in a world where simplicity, balance, harmony were being systematically erased by the Machine, cultural expression followed suit. Deformed by a colossal defeat, a revolt against the primitive and natural sets in. Industrial discipline—the latest and deepest form of domestication—infects all of society. Toynbee noted that "mechanization spelled regimentation...[which] had taken the spirit out of a Western industrial working class and a

---

4   Edward Carpenter, *Civilization—Its Cause and Cure* (London: S. Sonnschein, 1897), p. 3.

5   Renato Poggioli, *The Theory of the Avant-Garde* (Cambridge, MA: Belknap Press, 1968), p. 75.

Western middle class in succession."[6] Early on Stendhal saw that "one of the consequences of the modern dedication to productivity was sure to be the exhaustion of the natural human gift for the enjoyment of life."[7] Weariness of mind; potentially subversive energies suppressed.

The dominant minds—Comte, Darwin, Marx, etc.— agreed: the ascending order of civilization required always more complexity, homogeneity, work. In the 1880s Havelock Ellis recounted his "feeling that the universe was represented as a sort of factory filled with an inextricable web of wheels and looms and flying shuttles, in a deafening din."[8] The philosopher Arthur Schopenhauer did not at all join in the official optimism. His idea of pessimism, however—in view of the overall failure of desire and will—posited that will itself was the underlying problem. A classic case of deformed thinking.

Decadent literati in the West gave voice to a sense of nothingness at the heart of things.[9] In 1890 Max Nordau's very popular *Degeneration* depicted the fin-de-siècle mood as that of the impotent despair of a sick individual, dying by inches. The book is uneven, to say the least, but Nordau pointed accurately, in particular, to the nervous hysteria brought on by industrialization and the growth of cities. He wrote of an enormous increase in hysteria, and countless others concurred. Already in 1881, the French journalist Jules Claretie had declared, "The illness of our time is hysteria.

---

6   Arnold J. Toynbee, *A Study of History*, Volume IX (New York: Oxford University Press, 1954), p. 577.

7   Cesar Grana, *Fact and Symbol: Essays in the Sociology of Art and Literature* (New Brunswick, NJ: Transaction Publishers, 1994), pp. 169–170.

8   Karl Beckson, *London in the 1890s: A Cultural History* (New York: W.W. Norton, 1992), p. 318.

9   John A. Lester, Jr., *Journey Through Despair 1880–1914* (Princeton, NJ: Princeton University Press, 1968), p. 32.

One encounters it everywhere."[10] This was the paradigmatic psychological malady of late 19th-century Europe. Suicide rates rose to unprecedented levels, occasioning a considerable literature on the subject.[11] Suicide also became a common feature in fiction; Thomas Hardy's *Jude the Obscure* (1896) contains perhaps the most sensational fictional suicide of the era.

A proliferation of occult movements was another aspect of Decadent malaise, as was the rise in opium use. The strong popularity of Wagner's music, with its mythic religiosity, often assumed cultish proportions. The redemptive eroticism of operas like *Tristan und Isolde* provided a pseudo-utopian refuge from reality. The late-century rise of anti-Semitism, especially in Austria and France, was the disgrace of the century. Strange, even pathological, phenomena, in an ugly industrial world that is not being challenged.

Decadence is self-consciously artificial.[12] It bears an unmistakably anti-natural quality that is a sad reflection of the technological dominion that literally destroys nature. "My own experience," remarked Oscar Wilde in *The Artist as Critic*, "is that the more we study Art, the less we care for Nature."[13] The retreat into artifice, closing the door on the outside reality. As William Butler Yeats put it, referring to poetry: it is "an end in itself; it has nothing to do with thought, nothing to do with philosophy, nothing to do with life, nothing to do with anything...."[14]

---

10 Mark S. Micale, ed., *The Mind of Modernism* (Stanford, CA: Stanford University Press, 2004), Micale, "Decades of Hysteria in Fin-de-Siècle France," p. 84.

11 John Stokes, *In the Nineties* (Hemel Hemstead, UK: Harvester Wheatsheaf, 1989), pp. 121–122.

12 R.K.R. Thornton, *The Decadent Dilemma* (London, Edward Arnold, 1983), p. 32.

13 Elaine Showalter, *Sexual Anarchy: Gender and Culture at the Fin-de-Siècle* (New York: Viking, 1990), p. 170.

14 Beckson, *op. cit.*, p. 87.

Decadents saw a world in which survival meant keeping one's distance. With Symbolism, as the name proclaims, came a fuller retreat, into the strictly symbolic. Language as an independent quality takes over from meaning. This literary style is the effort of the Word to somehow express everything while confronting nothing. Meanwhile the International Date Line was established in 1884, a milestone of global, integrated industrialism. The Gatling gun lifted power imperialism to new heights, while skyscrapers and the Eiffel Tower (1889) showcased new vistas for the new order.

Outside of what we might call mainstream Decadence, however, there were some oppositional voices and actors. *News from Nowhere*, an 1890 novel by William Morris, depicts a harmonious, face-to-face world, devoid of factories. This utopian world of great beauty and humanness was a powerful response to Edward Bellamy's Marxist-oriented *Looking Backward* (1889). Morris rejected Bellamy's hymn to factories and regimented industrialism: "a machine life is the best which Bellamy can imagine for us on all sides; it is not be wondered at then, that his only idea of making labor tolerable is to decrease the amount of it by means of fresh and ever fresh developments in machinery."[15] How prescient Morris was. Well over a century later, it is easier still to see that work not only increases, but is steadily more alienated.

In France, Alfred Jarry, in his Ubu plays, also expressed antipathy to "machine life"—indeed to any routinized approach. In *Ubu Agog*, for example, "free men" in an anti-disciplinary army scrupulously disobey every order.[16] Jarry must have been quite aware of those who did not confine their anti-authoritarianism to the stage—the anarchists. In fact the anarchist upsurge of the early 1890s was a major

---

15 William Morris, "Looking Backward," *Commonweal*, 22 June 1989.

16 Roger Shattuck, *The Banquet Years: The Arts in France, 1885–1914* (New York: Harcourt Brace and Company, 1958), p. 227.

public preoccupation in France, featuring as it did a series of bombings.

The French working class had been decimated by the bloody repression of the Paris Commune in 1871, but by the mid-1880s intolerable conditions provoked more and more wildcat strikes and made anarchism appealing. Belgium, too, experienced similar developments, including the 1886 wave of vandalism and strikes in Liège, and bands of unemployed people roaming towns and countryside in the Meuse and Hainaut regions. Eleven anarchist bombs exploded in Paris between 1892 and 1894; French President Sadi Carnot was assassinated in 1894.

This decidedly non-Decadent aspect of fin-de-siècle Europe not only appealed to various workers, especially to artisans threatened by the ascendant industrializing order, but also to some of the intelligentsia. "Propaganda by the deed" reached its peak coincident with the mature phase of Symbolism, and some writers were won over to the cause. Several well-known painters also stood with the anarchists, including Georges Seurat, Paul Signac, Maximilien Luce, Camille and Lucien Pissarro, and others. Renato Poggioli even referred to the "alliance of political and artistic radicalism."[17]

Of course anarchists, no matter how militant, were not all opposed to the Machine, any more than were writers and artists as a group. Kropotkin, for instance, believed in the potential of modern technology and wholeheartedly accepted industrialization, the foundation of all modern technology. Henri Zisly spoke out for Nature and decried the industrial blight, but was definitely in the minority within anarchism. He was part of the naturist movement that emerged in the 1890s, but did not flourish.[18]

---

17 Poggioli, *op. cit.*, p. 11.

18 Anonymously edited, *Disruptive Elements: The Extremes of French Anarchism* (Berkeley, CA: Ardent Press, 2014), pp. 228–231.

France underwent fairly sudden, profound changes with the industrializing process, including a collision with its long-running craft tradition. The overall breakdown in craftsmanship reverberated in the cultural sphere and led, in desperation, to methodical, scientistic styles that resembled those of technology.[19] It was a time of endings, of the loss of long-established foundations. "Never before have so many artists and writers been so obsessed with various processes and manifestations of decay," according to David Weir.[20] The founding in 1886 of a literary and cultural journal called *Le Décadent* announced the arrival of a cult of highly self-conscious sensual and aesthetic decadence. The ethos is one of decreasing vitality and hopelessness,[21] of standing "in the presence of the dissolution of a civilization."[22]

Again, one could find contrary sentiments. Reviewing Jarry's *Ubu Roi*, Arthur Symons noted the "insolence with which a young writer mocks at civilization itself."[23] A mural by Paul Signac entitled "In the Time of Harmony: the Golden Age is Not in the Past, It is in the Future" (1893–1895) recalled Morris' *News from Nowhere* with its bucolic, anti-industrial pleasures. The primitivist paintings of Henri Rousseau and Paul Gauguin also come to mind.

Gustave Flaubert's *Madame Bovary* (1856) was an early, deadly satire of French Romanticism; Emma Bovary's suicide was a bitter commentary on the fruits of emerging consumerism. Naturalism's gritty realism was a tableau of

---

19 Wylie Sypher, *Literature and Technology* (New York: Random House, 1968), p. 73.

20 David Weir, *Decadence and the Making of Modernism* (Amherst, MA: University of Massachusetts Press, 1995), p. xii.

21 William Barry, *Heralds of Revolt: Studies in Modern Literature and Dogma* (London: Hodder and Stoughton, 1904), p. 215.

22 Arnold Hauser, *The Social History of Art*, Volume Four (New York: Vintage Books, 1958), p. 185.

23 Beckson, *op. cit.*, p. 336.

subjugation; Emile Zola's *Germinal* (1885), for example, a tale of misery and hopelessness.

Joris-Karl Huysmans provided in 1884 what has been widely called the bible of Decadence with his novel *À Rebours*, usually translated as *Against the Grain* or *Against Nature*. Here nature has truly been left behind by an over-civilized seeker of rare sensations, the protagonist the Duc des Esseintes. His escape into artifice ends in ridiculous exhaustion as he finally seeks a consoling faith. The full range of Decadent sensibilities are displayed or predicted.

One can notice that in Claude Monet's paintings after 1883, the human figure shows up less and less and finally disappears. Something similar is happening with the new Symbolist poetry, championed by Paul Verlaine, Stéphane Mallarmé and others. In fact, Symbolism and Decadence became more or less synonymous in the 1890s.

The essence of Mallarmé's impersonal syntactical mannerisms is revealed by his description of a dancer. He said she is not a woman who dances, nor even a woman, but a metaphor; that is, it requires a poet to make the dancer real.[24] Mallarmé realized that there can be no stable art, no classical forms, in an unstable society. From this he derived the axiom that all great poetry must be incomprehensible. Arthur Rimbaud's 1870s work pioneered what came to be called Symbolism, but R.C. Kuhn's assessment seems valid: "Rimbaud's poetry is a rejoicing in presence; Mallarmé's is a celebration of absence."[25] Borne along by a current of language and very little else.

Many of these writers ended up becoming everything they once abhorred. Rimbaud was a gun-runner in Africa,

---

24   Richard Candida Smith, *Mallarmé's Children* (Berkeley, CA: University of California Press, 1999), p. 71.

25   Reinhard Clifford Kuhn, *The Demon of Noontide: Ennui in Western Literature* (Princeton, NJ: Princeton University Press, 1976), p. 317.

Jarry and Verlaine died of alcohol, Huysmans died a Catholic—a litany of failed and foreshortened lives. The ugly anti-Semitism of the Dreyfus Affair from 1898 onward marked the end of the period of Decadence in France. Before long a healthier, combative era in culture began. We may say this change was already in the air when a student riot erupted without warning in Paris in June 1893. An art-ball crowd of painters, poets and the like became an unstoppable force, occupying the Latin Quarter and requiring no fewer than 30,000 troops to disperse.[26]

English Decadence, though generally a bit less hard-core, drew a lot from French models and precepts. Less absolute than the French but with the same lack of interest in life and action, the same sense of the futility of it all. John Ruskin, who like William Morris after him championed craftsman-ship, saw in the 1860s that "progress and decline" were "strangely mixed in the modern mind."[27] In 1893 Arthur Symons described Decadence as a "beautiful and interesting disease."[28] Gossip and its enactment, scandal, were symptomatic. Decadents seemed titillated, even seduced, by the idea of corruption. The word "morbid" became something of a cliché by the end of the century.

The earlier cultural synthesis of Victorianism was unraveling in an ethos of exhaustion and pointlessness. In Wilde's *The Picture of Dorian Gray* is the following all-too-resonant exchange:

"Fin de siècle," murmured Lord Henry.

"Fin du globe," answered his hostess.

---

26   Cesar Grana, *On Bohemia* (New Brunswick, NJ: Transaction Publishers, 1980), pp. 374–375.

27   J. Edward Chamberlin and Sander L. Gilman, eds., *Degeneration: The Dark Side of Progress* (New York: Columbia University Press, 1985), Sandra Siegel, "Literature and Degeneration: The Representation of Decadence," p. 199.

28   Beckson, *op. cit.*, p. xii.

"I wish it were fin du globe," said Dorian with a sigh. "Life is such a great disappointment."[29]

The disintegration of a high Victorian ideal of English civilization bore the usual marks of increasing mechanization, notably greater nervous exhaustion from a more intense pace of life.[30] A boom in interest in the occult, more drug use, the usual Decadent-era escape routes. And in response, new efforts at social integration, like more compulsory education and a bigger emphasis on organized sports. Meanwhile, Decadents pursued their perverse and escapist paths. They saw ugly industrial urbanization—and embraced the city as the supreme work of artifice. They embraced what they saw as inescapable rather than try to oppose it. It is ironic, in an age of irony, that world-weary and ennui-filled Decadents were often obsessively drawn to the vitality of working-class pubs and music halls.[31] But the typical Decadent poet, Ernest Dowson, does not appear to have made even vicarious use of such vitality. Bored to death by the nothingness of everything, his lines seem to almost always end on a note of disillusionment. The Pre-Raphaelite art of Gabriel Rossetti and others in the 1850s, in its Ruskin-like distaste for industrialism, was something of an influence much closer to the end of the century. But its subjects appear flat and doll-like, depthless—qualities that generally fit the Decadent profile.

Elaine Showalter has explored what she called the sexual anarchy of fin-de-siècle England, in particular the threat of feminism to a very sexist culture. Robert Louis Stevenson's

---

29 Oscar Wilde, *The Complete Works of Oscar Wilde*, Vol. 3 (New York: Oxford University Press, 2005), *The Picture of Dorian Gray*, p. 318.

30 Peter N. Stearns, *The Industrial Revolution in World History* (Boulder, CO: Westview Press, 2007), p. 172.

31 Mikulas Teich and Ray Porter, eds., *Fin-de-Siècle and its Legacy* (New York: Cambridge University Press, 1993), Alison Hennegan, "Aspects of Literature and Life in England," p. 197.

The Strange Case of Dr. Jekyll and Mr. Hyde (1886) as a myth of warning to women of the dangers outside the home, also a case study of male hysteria and homophobic panic; Bram Stoker's *Dracula* (1897) as a fantasy of reproduction through transfusion, that is, without the need of women.[32] The Sherlock Holmes figure is also of interest; he turns to cocaine out of his ennui and boredom. The Arts and Crafts movement of the 1880s, itself an outgrowth of Pre-Raphaelite sensibilities, failed to gain traction. Its key figure, William Morris, was reduced to designing wallpaper and furniture for the rich, which he privately called "rubbish."[33] A rare sign of life was *Jude the Obscure*, Thomas Hardy's last novel. Influenced by the French utopian Charles Fourier, it contains very explicit social criticism and an early ecological awareness.

Weak, low-energy Decadence had little with which to sustain itself. Despite its showy and sometimes shocking bohemianism, several prominent Decadents retreated into the Catholic church: the artist Aubrey Beardsley, Oscar Wilde, and the poets Ernest Dowson, Lionel Johnson, and John Gray. Beardsley died of tuberculosis in 1898 (at age 25), as did Dowson in 1900. Critic Arthur Symons suffered a mental breakdown in 1908, and poet John Davidson commited suicide in 1909. Oscar Wilde, who did sense the underlying rot of civilization, died in 1900, which was already just past the sell-by date of Decadence in England.

"Vienna in the fin de siècle [experienced] acutely felt tremors of social and political disintegration."[34] Receptive to ideas of Decadence elsewhere, the Hapsburg Empire capital exhibited ever stronger symptoms of decline. Writing of

---

32 Showalter, *op. cit.*, pp. 127, 107, 179.

33 E.P. Thompson, *William Morris* (New York: Pantheon Books, 1977), p. 109.

34 Carl E. Schorske, *Fin-de-Siècle Vienna: Politics and Culture* (New York: Alfred A. Knopf, 1980), p. xvii.

1890s Austria, Robert Musil recalled a sense that "time was moving faster than a cavalry camel.... But in those days, no one knew what it was moving towards."[35] Even more than in France, the pace of industrialization was intense and disruptive, the "new conditions of modern life emerging suddenly and uncontrollably."[36] Czech critic Frantisek Salda characterized 1890s Vienna as a culture in which "young men imitated old with their tiredness, wornness, blague and cynicism."[37] Progress as a positive thing seemed at an end.

This deflation or defeat, again, had a deeper basis. Life on a human scale was being erased in society at large. Frederick Morton referred bitterly to the "industrial flowering" and its effects on the worker, who before "served the needs of specific men. Now he was a nameless lackey to faceless machines."[38] Along with Europe's highest suicide rate came well-trodden Decadent dodges: avoidance of socio-political reality; an occult revival; the elevation of subjectivism; Wagner worship with its ersatz pietism, pseudo-redemption, and virulent anti-Semitism; embrace of Schopenhauerian pessimism/ nihilism; aversion to nature.

In terms of subjectivism or inwardness, the emergence of Sigmund Freud fits the overall predicament. 1890s Vienna saw his metaphysic mature. At base, Freud's analysis rules out the relevance of any politics in favor of the primacy of very early sexual development and the primal conflict between father and son. Certainly no decadent, Freud was nonetheless part of the retreat from outside reality.

---

35   *Ibid.*, p. 116.

36   Micale, *op. cit.*, p. 88.

37   Robert B. Pynsent, ed., *Decadence and Innovation: Austro-Hungarian Life and Art at the Turn of the Century* (London: Weidenfeld and Nicolson, 1989), Magda Czigany, "Imitation or Inspiration," p. 119.

38   Frederick Morton, *A Nervous Splendor: Vienna 1888/1889* (Boston: Little, Brown, 1979), p. 314.

Another precursor of modernism was Robert Musil, whose *The Man Without Qualities*, while not published until 1930, was set in Decadent Austria. The characters in the novel search for order and meaning in a culture which has broken down into a state of spiritual crisis. The sense of a loss of reality is paramount, and although Musil is not explicitly interested in social specifics, he invokes the slide-rule as a reigning symbol, not unlike the computer today. Mainly we see the turn toward language, away from the moral standstill at large, soon to be so greatly stressed by Wittgenstein and others. Musil's hero Ulrich is indeed "without qualities." His character dissolves into a multiplicity of divergent, even opposing selves. The non-coherence of the modern mind is another feature, along with Musil's stress on the merely linguistic, previewing postmodernism a century later.

What could be termed gigantism in serious Viennese music echoed enormous factory growth at this time. Gustav Mahler, key composer and conductor, orchestrated long symphonies for one hundred or more players, often accompanied by huge choruses.

Gustav Klimt led the art nouveau Secession movement, but this artist-heretic "quickly acquitted strong social and financial backing."[39] Modern art, somewhat ironically, came into official favor just when parliamentary government was virtually falling apart, largely because of the poisonous rise of anti-semitism early in the 1890s.

In Germany, too, pessimism led to the cultivation of aestheticism as avoidance. The novels of the 1890s are devoid of realist content, and the major poets (e.g., Stefan George, Rainer Maria Rilke, Hans von Hofmannsthal) likewise refrained from dealing with the world, in favor of giving voice to fleeting impressions, moods, and perceptions. A

---

39   Schorske, *op. cit.*, p. 208.

spate of plays and novels, however, depicted how German secondary education produced adolescent misery, including suicides.[40] As in most countries, industrialization increased inequalities of wealth and income, while tuberculosis was a scourge in Berlin and other cities.

The air of unreality was also felt by Czechs as rapid industrialization swept away most of the past. Arthur Breisky described the dandy who "is the knight of todays; he closes his eyes indifferently to all tomorrows."[41]

In Hungary, poet Gyula Reviczky decided that "the world is but a mood,"[42] in step with the hopelessness and flight from society of Decadence in the rest of Europe. But Endre Ady, who started a new epoch in his country's literature, was a fine counter-example. He was a radical anti-feudal social critic who attacked the values of work and efficiency, and advocated simplicity and beauty. A definite non-embrace of the Machine.

And our own period of Decadence? Are we not more "over-civilized" than ever, in greater denial? There is more of the artificial than before, and an even greater indifference to history. Our sense of hopelessness is profound, a techno-industrial fatalism: the inevitability of it all. In 1951 Karl Jaspers wrote of "a dread of life perhaps unparalleled" as modernity's "sinister companion."[43] "As mechanization takes place…man loses his way amid the growth of complexity; he loses the sense of reality, of his own personality."[44] In our own age, Frederic Jameson points to a general "waning

40   John Neubauer, *The Fin-de-Siècle Culture of Adolescence* (New Haven, CT: Yale University Press, 1992), p. 2.

41   Pynsent, *op. cit.*, "Conclusory Essay," p. 177.

42   *Ibid.*, p. 143.

43   Karl Jaspers, *Man in the Modern Age* (London: Routledge & Kegan Paul, 1951), p. 62.

44   *Ibid.*, p. 169.

of affect,"[45] the cumulative impact of Progress at the expense of affective, or emotional life.

Nothing could be more obvious than that the eco-disasters of Decadence this time are industrially produced. Flattened, bored, deskilled personal lives find their double in the decimated, impoverished physical world. As Jaspers summed it up, "The machine in its effect upon life and as a model for the whole of existence."[46]

A retreat to aestheticism can be no resolution to what can only be fully faced outside of the aesthetic realm. Freud was right in pointing out that art is not a pleasure but a substitute for pleasure.[47] A complete life would not require the consolation of art.

Edward Carpenter looked at civilization as a kind of disease we have to pass through.[48] This Decadence can be overcome. Confronting the nature of the whole is the inescapable challenge.

45   Frederic Jameson, *Postmodernism or the Cultural Logic of Late Capitalism* (New York: Verso, 1991), passim.

46   Karl Jaspers, *The Origin and Goal of History* (New Haven, CT: Yale University Press, 1953), p. 144.

47   Sypher, *op. cit.*, p. 203.

48   Carpenter, *op. cit.*, p. 1.

CONCLUDING ANTI-HISTORY POSTSCRIPT

We know that the past is always molded to sanction the approved order, its government and social institutions. History is written by the victors. The story of civilization is not told by those who lost to the domesticators, the civilizers. But as we have seen, recurring transitions and crises are proof that civilization never enjoys a long, untroubled sleep.

Its ideologues have always presented a different picture, one of stability and pacification. A famous somewhat recent example is Francis Fukuyama's *The End of History and the Last Man* (1992), announcing the victorious end to the evolution of civilization. The world system of capital and technology is complete, upon the end of the Cold War; no further rough seas to cross. But in less than a decade the Anti-Globalization movement (1999–2001) provided a strong challenge to that hegemony in North America and Europe.

There's a lot more to history than questions of accuracy, of fidelity to events and currents. The most basic question would seem to be: What is history? In James Joyce's *Ulysses*, Stephan Dedalus says, "History is a nightmare from which I am trying to awake." Theodor Adorno referred to "the infernal machine that is history," pointing to its continuum of suffering. Everything has a history, and history has everything. Domestication requires storage; history is a form of storage.

Walter Benjamin counseled that we must go against the historical movement. The limits of history are increasingly being revealed to us. The historic dimension wears the mask of death. If the past is somehow to be redeemed, that redemption will occur outside of history.

Historiography, the study of history, does not concern itself with time. But the nature of history is very deeply tied to the question of time, the regime of time, its ever-greater materiality and oppressiveness. The continuity of history—and time—is imposed and alienating. Time is more than a medium; like technology, it is far from neutral.

In the 1980s I came upon a passage in Walter Benjamin's "Theses on the Philosophy of History," and was immediately intrigued by this now well-known piece. It is his meditation on a 1920 painting by Paul Klee, *Angelus Novus*:

> Where we perceive a chain of events, he sees one single catastrophe which keeps piling wreckage upon wreckage and hurls it in front of his feet. The angel would like to stay, awaken the dead, and make whole what has been smashed. But a storm is blowing from Paradise; it has got caught in his wings with such violence that the angel can no longer close them. This storm irresistibly propels him into the future to which his back is turned, while the pile of debris before him grows skyward. This storm is what we call progress.
>
> —Walter Benjamin, *Theses on the Philosophy of History* (1940)

Benjamin's interpretation of Klee's angel seems to me profoundly insightful. The storm blowing from Paradise is time, which becomes history and progress. The pile of debris is the course of civilization, growing skyward.

The book you are now reading is a testimony to the need for historical awareness; but Benjamin points us further. A messianic dimension is needed if history is to be redeemed, if a part of the happiness our ancestors could not have is to be validated. To "awaken the dead, and make whole what has been smashed." To unmask the paradigm of history and its fundamentally legitimating enterprise.

Outside the symbolic system, beyond representation; what Lacan calls the encounter with the Real. Time and history ceaselessly advance all-encompassing domination; so a rupture, a break is needed. Only then could humanity realize a past, citable in all its lived moments, un-reified.

This vision is the opposite of Hegel's totalizing notion of history as the process by which the principle of freedom actualizes itself. Breaking the spell in a frankly apocalyptic way is Benjamin's counter-offer. A glimpse of this was presented in 1830, when radicals fired at clock towers.

Benjamin provides a striking contrast with the promise of historical advancement:

> Marx says that revolutions are the locomotive of world history. But perhaps it is quite otherwise. Perhaps revolutions are [or should be] an attempt by passengers on this train—namely, the human race—to activate the emergency brake.

The brake is a break with history. We were conscripted into history and we must make our exit from it.

DONE IN FROM WITHIN: AN INTEMPERATE REVIEW

All the leftist parties in the West took sides in the four-year industrial slaughterhouse of World War I. Leading anarchist Peter Kropotkin also supported it.

Emma Goldman was deported to Russia in 1919 and remained there until 1921. She was a partisan of the Bolsheviks' takeover of the revolution even after Lenin told her he had no intention of altering his policy of murdering and imprisoning anarchists. It was not until Lenin and Trotsky massacred the insurgent sailors and anarchists of the Kronstadt naval base in 1921 that she finally broke with the red fascist Bolsheviks.

Writers sympathetic to Goldman refer to her disillusionment with Soviet rule, but tend to gloss over her outright complicity.

In 1936, popular front governments came to power in France and Spain. These leftists regimes reinforced industrial work discipline just as much as their predecessors. An indispensable book on the subject is Michael Seidman's *Workers Against Work: Labor in Paris and Barcelona During the Popular Fronts* (1991). The nationally supervised efforts of output and productivity were fully endorsed by the anarcho-syndicalists, who were all about mass production, like the other left-wing elements. This partnership was further cemented in Spain when the anarcho-syndicalist union, CNT, joined the government.

Skipping forward to 1999–2001, we come to the so-called Anti-Globalization movement. So-called because few of its participants questioned globalization itself, or indicted the globally integrated web of civilization. The movement was kicked off in 1999 with an anarchist riot in Eugene, Oregon and then, more significantly, in Seattle near the end of the year. A World Trade Organization summit was prevented from happening, and some of downtown Seattle was trashed. Globalization summits in Quebec City, Prague, and Genoa in 2000 and 2001 attracted large and vigorous protests. 300,000 took to the streets of Genoa in July 2001, and a young anarchist was killed by police. A major clash, and very revealing.

At one point the police forces attacked the large Black Bloc contingent, dividing it into two parts and driving much of it down toward the beach area, which was a home base for many of the protestors. World Social Forum people, Chomsky-type leftists, did their best to bar the gate to the beach, so that the radicals were trapped by the police assault. The WSF organization later slandered the courageous Black Bloc fighters as agents provocateurs.

The Zapatista movement in Chiapas, Mexico drew much-deserved interest and support. Its famous Sub-comandante Marcos declared that he had been a leftist urban intellectual but had come to see indigeneity as the heart of liberatory politics: respect for the land, for community. But the EZLN's Sixth Declaration of the Lacondon Jungle changed all that in 2005. It called for a new political party and launched the "Other Campaign" in 2006. The new policy was for "good governance councils," a new national constitution, citizenship, and the EZLN was soon running a candidate for president of Mexico. So much for indigeneity! And a strange (not so strange, sadly) silence then descended with this turn to the Left on the part of anarchists and others.

The 2011–2012 Occupy movement was a reformist social justice phenomenon, led by progressives like David Graeber. Lacking any radical orientation, it was largely ignored by anarchists. Its most militant expression was Occupy Oakland, which managed a one-day closure of the Port of Oakland.

De-occupy or de-colonize would have been at least the first step in a radical direction, but this was explicitly refused in favor of a vote celebrating "the legacy of the Left." Pathetic, and characteristic of the liberal-left nature of Occupy.

2016–2017 saw massive resistance to the Dakota Access Pipeline on the Standing Rock Sioux Reservation in North Dakota. Thousands joined the Sioux and had supporters worldwide, as skirmishes and standoffs took place for several months, even through a bitter high prairie winter. Undefeated by military force, the effort to prevent the pipeline was rather suddenly ended by order of the Sioux chiefdom. This critical, fatal development was largely ignored.

So often, the failure comes from within, a more or less "fifth column" phenomenon. Often the undermining comes from the Left, and the lesson has yet to be fully learned, it seems.

A NOTE ON FREEDOM

In any but a personal sense, the word freedom now has a quaint ring to it. The individual clings to a notion of freedom in terms of private autonomy, success, fulfillment, against a backdrop of surveillance and political disempowerment. We are so much data, interpreted and manipulated by algorithms, not to mention the political racket.

Freedom, to some at least, was once a fiercely contested word. Who can get excited about it today, in mass society–which has erased freedom about as surely as it has erased community. We have been integrated into the global machine of technology and capital, a process made to seem inevitable, and therefore irresistible. Modernity undermined and all but put an end to concepts of freedom and autonomy. How to realize freedom had been at the forefront of political philosophy in the West for centuries? Within civilization it has never been a reality.

But this is far from the whole picture. As Fredy Perlman, Kevin Tucker, and many others have stressed, the fight to be free is always alive, even under the most negative circumstances.

Let me cite Jean-François Steiner's *Treblinka*[1], a book that has had a profound influence on me since I first read it in the 1960s. Steiner was a prisoner at the eponymous death camp, and tells an unfolding tale of the struggle for liberation there—an epic lesson, it has seemed to me. It is the mostly Jewish inmates who are forced to the work of Treblinka's murder facility, and as the story begins they are so dehumanized as to be severely unproductive. Their overseers must issue separate commands for each step of a given task, such that the overall operation is barely functional. And when

---

1   Jean-François Steiner, *Treblinka* (New York: Penguin Random House, 1994).

they necessarily reduced the brutal pressure on the inmates, the first free outcome was a spate of suicides. It was from this point on that the idea of resistance grew, and painful, usually unsuccessful, steps were taken. Ultimately, rather unbelievably, six hundred prisoners took over, killing their Nazi captors, burning several crematoria, and escaping into the woods. A true story about freedom prevailing against a maximum of oppression.

It may be the case that we are free to the extent the dominant order allows for its successful functioning. But Treblinka reminds us that it is also enough for resistance to grow and even blossom.

How do we confront the basic restrictions on the freedom of even the freest in an unfree society? What can it mean to speak of freedom itself in the context of the alienated freedom of the technosphere which is late civilization?

The responses have been varied, to say the least, and ultimately fruitless. Epicurus espoused free choice and spontaneity. Kierkegaard saw anxiety and bewilderment from a sense of freedom, Freud said we are governed by unconscious thoughts and feelings, and so on. Enrico Manicardi points out that civilization not only impinges on freedom, but reduces the desire for such an environment, largely replacing it with fear of freedom.[2]

An insidious progression, that of every deepening domestication, has made talk of freedom basically meaningless and irrelevant. Insofar as domestication and its spawn, civilization, are unquestioned, freedom is off the table.

The entire philosophical tradition of the West, especially since Spinoza, Leibniz, and Kant, has equated free activity with activity in line with reason. As if reason were a well-understood given and not in service, non-neutrally, to the fundamental

---

2   Enrico Manicardi, *Free From Civilization* (Berkeley, CA: Green Anarchy Press, 2012), pp. 174–175.

order. The meaning of reason, what is rational, is what reigns. We are free to the extent, that is, that we are unfree.

The Grand Inquisitor chapter of Dostoevsky's *The Brothers Karamazov*[3] is the quintessential encounter between authority and freedom. Jesus reappears, but is rejected in favor of the security of the Church, its rules and structure. In a similar vein, anthropologist Franz Boas, stressing the subjective part of freedom, stated that "A person who is in complete harmony with his culture feels free. He accepts voluntarily the demands made upon him." Bronislaw Malinowski responded that by Boas' standard, one who accepted Nazi ideology was thereby free.[4] There is indeed the "freedom" to adapt and conform. Nietzsche's concept of loving one's fate (*amor fati*) is acceptance, not freedom; Sartre's supposedly radical idea of choice is a similar demand to embrace the world as it is.[5]

The process of colonization and Westernization has employed an idea of freedom as ideological assistance to its project of unfreedom. Koshone Mahbubani is among those who therefore advocate dewesternizing freedom.[6] Despite genocide against indigenous people and enslavement of Blacks, U.S. identity has been based on a fundamentalism of liberty. "Land of the free," even to the extent of belief in the founding colonizers' greatest freedom of all: freedom from history.[7]

This absurd view has faded, and the ideal of actual freedom isn't faring very well either. There is some consensus

---

3   Fyodor Dostoevsky, *The Brothers Karamazov* (New York: Modern Library, 1996).

4   Moises Lino e Silva and Huon Wardle, editors, *Freedom in Practice* (New York: Routledge, 2017), Introduction.

5   Anthony Farr, *Sartre's Radicalism and Oakeshott's Conservatism: The Duplicity of Freedom* (New York: St. Martin's Press, 1998), p. 86.

6   Walter D. Mognolo, *The Darker Side of Western Modernity* (Durham, NC: Duke University Press, 2011), pp. 297–298.

7   Eric Foner, *The Story of American Freedom* (New York: W.W. Norton, 1998), p. 332.

that the struggles of the 1960s could be characterized as fights for freedom at base. Since then, however, the view has obviously darkened.

Rethinking Freedom: Why Freedom Has Lost Its Meaning and What Can Be Done To Save It is the ambitious title of C. Fred Alford's effort.[8] Unfortunately, Alford pretty much comes up empty. His personal survey of students finds a lot of ironic detachment and lack of interest in the topic. A "tone of dismissal, often bordering on contempt," was a common response, he found.[9]

William Butler Yeats summed it up well:

> "We have fed the heart on fantasies
> The heart has grown brutal from the fare."[10]

Postmodern culture has undermined the idea of emancipation. In an age of unmatched integration of the subject with the totality, postmodernism valorizes the free-floating individual, unmoored from all ties, from what is left of social solidarity. Freedom from the land, from all species, including our own. In typical postmodern jargon/nonsense, Jean-Luc Nancy states that freedom is groundlessness, the "experience of experience," an experience of surprise at an existence that appears from no place and means nothing.[11] This is the perspective of evasion, the voice of an ethos of no resistance.

Freedom means resistance, facing up to the need to destroy what constitutes and enforces the absence of freedom. It is

---

8   C. Fred Alford, *Rethinking Freedom: Why Freedom Has Lost Its Meaning and What Can Be Done to Save It* (New York: Palgrave Macmillan, 2005).

9   *Ibid.*, p. 2.

10   William Butler Yeats, "The Stare's Nest by My Window," in *The Tower* (New York: Scribner, 2004).

11   Jean-Luc Nancy, *The Experience of Freedom* (Stanford, CA: Stanford University Press, 1993), pp. xxii–xxiii.

transgressive, aimed at all the institutions that make up taken-for-granted captivity. It flourishes only in opposition to our conditioning, to our tacit acceptance of what is. As Thoreau had it, "The greater part of what my neighbors call good I believe in my soul to be bad, and if I repent of anything, it is very likely to be my good behavior."[12]

---

12 Henry David Thoreau, *Walden* (New York: Heritage Press, 1939), p. 20.

ACTUAL NIHILISM: SOME THOUGHTS ON THE SF BAY AREA IN THE SEVENTIES

In 1966 I went from Stanford to Haight-Ashbury, and remained in San Francisco or Berkeley for the rest of the '60s. What started in the Bay Area in 1964 with the Free Speech Movement ended there at close of decade. A critical point was reached in Berkeley, the fight to defend People's Park, a small liberated space that creatively sprang up just off Telegraph Avenue in the spring of 1969. On May 15, a march down Telegraph was fired on by cops, one dead, scores wounded, including one blinded by shotgun fire. This was the "bloodbath" California Governor Ronald Reagan had urged. Riots raged during Berkeley nights for weeks afterward, but this was to be the extended last hurrah, it turned out.

The Movement of the '60s, as it is known, seemed to come to a sudden end. All the air went out of the balloon overnight, or so I experienced it. Everything that might have been projected was no more, and everyone apparently knew it. No one, say, called a meeting, because no one would come.

The violent debacle east of Berkeley, where the Hells Angels killed someone during the Rolling Stones concert, was a sadly fitting marker of the end. The years of hope and resistance thus ended at Altamont, December 1969. The two exceptions to this were the radical prisoners' element emerging at the end of the decade, and the radical women's movement that took shape in the early '70s.

These efforts came too late to give life to the general temper of the times, and had to try to go forth in the absence of the wider coalescence of social movements that made up "the '60s." Which is not to say that the fire went out completely, rather that defiant acts were basically isolated and desperate, no larger current behind them.

For example, the Black Panther Party, founded in Oakland in 1966, endured police killings and internal division and was in decline after 1968. Bombings of corporate and government offices by the Weathermen/Weather Underground continued throughout 1970 by its surviving members, now fugitives. The 19-month takeover of Alcatraz in San Francisco Bay by Indians of All Tribes ended in June 1971.

Many had a very hard time coming to grips with the evident end of possibilities for the future. The original innocence of, say, the Haight-Ashbury hippie scene had long since descended into a hard drugs ambience, a shocking development. The fallout in terms of pop culture was depressingly clear. Jimi Hendrix and Janis Joplin were dope fatalities in 1970; Jim Morrison, hounded out of the U.S., died in Paris in 1972.

The sunny, upbeat harmonies of the Beach Boys were besieged by leader Brian Wilson's mental health crises and by the premature death of two of its members. The weariness of their 1973 single "Sail On Sailor" was a sign of the times. The Carpenter duo, siblings Richard and Karen, was swamped by his addiction to Quaaludes and her fatal anorexia. Their sad reality gave the lie to the All-American, wholesome, clean-cut conformist message of their music.

August 1971 saw a bloody and unsuccessful attempt at the Marin County Courthouse to free George Jackson from San Quentin State Prison. His death provoked violent attacks on cops and police helicopters. The Attica prison uprising happened in September; the takeover and standoff ended with 43 dead. Both hostages and prisoners were almost entirely killed by the invading pigs.

In 1973 the Symbionese Liberation Army formed. About a dozen in all, mostly women and Blacks. They assassinated the Oakland superintendent of schools, kidnapped Patty Hearst, robbed banks, and ended up dead or in prison. They had to have known that the Movement was long dead, but jumped off into action anyway.

Some of us discovered Situationist ideas and small groups at the beginning of the decade. A somewhat anti-authoritarian critique of the Left was of interest, but the heyday of the Sits was the "French May" of 1968, an increasingly distant memory, it seemed.

The nihilist energy of punk, an import from England in late 1976, dominated the San Francisco subculture in '77. Its explosive anger spawned dozens of local bands, e.g., Negative Trend, UXA, Avengers, Crime. The biggest venue was Mabuhay Gardens, a former Filipino supper club on Broadway. Occasionally a carload of punkers would drive straight up Broadway to Pacific Heights, jump out, and smash the hell out of posh cars up there.

But the snarling negativity of the music was more backgrounded by heroin than by blows against the dominant order. Subsequent years of punk were even less likely to exhibit radical activity.

Defeat was the obvious reality as any semblance of social movements ebbed away. The '80s, of course, were to be even worse. By the end of the '70s I felt grief and anger with no outlet, sunk into heavy drinking, writing very little.

The tension between what, in its best moments, had been, and what was left was a tableau of nihilism.

ALSO A SPIRITUAL MOVEMENT

At a talk I gave in Turkey some years ago, a young woman said that the green anarchy phenomenon is at base a spiritual movement. Extremely intrigued—and surprised—by her judgment, I wanted to hear more. She had to catch a bus for home so this was not to be, but this view has resonated with me ever since. And I'm certainly not alone. "I think green anarchism is unique in that it finds importance in seeking out some form of spiritual connection," was an anonymous recent comment, a commonly made point.

Arguably, green anarchy/anti-civilization/anarcho-primitivist perspectives more than seek out "some form of spiritual connection"; they pretty much are that connection. Sometimes referred to as primal anarchy, it is, in my opinion, in its depth and the depth of its yearning for an altogether different world, that its basics are revealed. Basics, I think, that are rightfully termed spiritual.

To me, the primary aspects include wholeness, which means, among other things, as much distance from division of labor as possible. Presence or immediacy, unmediated life, relate to wholeness. Also simplicity, again related to the other values.

Intimacy with nature can be counted as spiritual, I would say; also wildness, the freedom of that which is most alive. So much that is there, embedded in the world beneath all that has been put upon it. What is still presented to us in no need of re-presentation. "Er gibt"—the world gives itself, in Heidegger's words.[1]

For many of us, it is the indigenous dimension, past and present, that informs and inspires us. A real link to all we have left. The idea of wholeness and connection is often

---

1   Quoted in Richard Wright, *The Heart of Wisdom: A Philosophy of Spiritual Life* (Lanham, MD: Rowman & Littlefield, 2013), p. 56.

called spirit by Native people. The association with breath and life is a hallmark of spirit, found in many times and places. Spiritual sites were of course wild.

Indigenous spiritual practices are land-based. As Gonzales and Nelson point out, "Land is everything to Native American peoples for several reasons."[2] This is an earth-grounded spirituality that does not lose connection to the land, and is a bond with all life on the land. For example, "For the Yupiit, society included both human and nonhuman members."[3] A sense of oneness, kinship, that is deeply spiritual and in which private ownership of the earth has no place.

We know what put an end to that reality and what has been the result. Jacques Monod referred to the "end to the ancient covenant between man and nature, leaving nothing in its place of that precious bond but an anxious quest in a frozen universe of solitude."[4] Lame Deer concluded that "Only human beings have come to the point where they no longer know why they exist. They...have forgotten the secret knowledge of their bodies, their senses, their dreams."[5] Communion with the world was lost, and so very much with it.

Many have understood this and have tried, in various ways, to find the spiritual within the barren cultures of civilization. Ralph Waldo Emerson, Thoreau's friend, was a pantheist transcendentalist who grasped the truth of a spirit-infused nature. Gary Snyder's *The Practice of the Wild* (1990) summed

---

2   Tirso A. Gonzales and Melissa K. Nelson, "Contemporary Native American Responses to Environmental Threats in Indian Country," in John A. Grim, editor, *Indigenous Traditions and Ecology* (Cambridge, MA: Harvard University Press, 2001), p. 499.

3   Ann Fienup-Riordan, "A Guest on the Table: Ecology from the Yupiik Eskimo Point of View," in Grim, *op. cit.*, p. 543.

4   Quoted in Gary Snyder, *The Practice of the Wild* (New York: North Point Press, 1990), p. 194.

5   Lame Deer, *Lame Deer, Seeker of Visions* (New York: Simon and Schuster, 1972), p. 157.

up his Buddhist reverence for undomesticated lands. In Taoism, an alternative to state-oriented Confucianism, many green anarchists have found anti-authoritarian sources, a clear spiritual-political crossover. "Reclaiming the Tao Te Ching for Anarchy" (*Green Anarchy* 17, Spring 2004) and "Reclaiming Chuang Tzu for Anarchy" (*Green Anarchy* 18, Summer 2004) exemplify this primitivist connection to the Tao.

Zen Buddhism emphasizes being mindful in the world at every moment, not fleeing into abstraction or the supernatural. Deep Ecology, first formulated by Arne Naess in 1973, articulates a spiritual vision of nature. But in my opinion, it remains too abstract to be either very spiritual or political. There are a host of nods to a spiritual-political orientation, but many of them, such as "sustainable spirituality" and "green Buddhism" seem faddish and lacking in substance. Spiritual tourism bespeaks the popularity of some forms of spirituality, which can be flexible and thin, lacking a cultural context.

Enter religion. Enter the categories of the sacred, the supernatural. Upon our separation from the natural world, an impulse arises to regain, or even surpass that previous connection. The word religion derives from the Latin *religare*, to re-tie, to heal the broken bond. Religion arises from need, the desire to re-establish what was lost.

So often it is approached via a common but entirely false assumption. "Religion is verily a universal feature of human culture," pronounced Robert Lowie.[6] "No society known to anthropology or history is devoid of what reasonable observers would agree is religion," saith Roy Rappaport.[7] "It is something that we have always done," according to historian of religion Karen Armstrong.[8] "Always done,"

---

6   Robert H. Lowie, *Primitive Religion* (New York: Boni and Liveright, 1948), p. xvi.

7   Quoted in Nicolas Wade, *The Faith Instinct* (New York: The Penguin Press, 2009), p 4.

8   Quoted in *ibid.*, p. 3.

unless one takes into account about 99 percent of our time as a species. Religion is absent among our non-domesticated hunter-gatherer forebears. Society was not divided into sacred and secular parts until relatively recently, less than 10,000 years ago.

Religion has greased the wheels of civilization, making palatable or at least softening the blow of the hard labor involved in the shift to domestication/agriculture. In fact, religion became a defining factor of the world's major civilizations. How is order to be kept? As the individual is diminished in civilization, religion advances, in inverse proportion. From no gods, to multiple gods, to the one Supreme Being god of monotheism, in Judaism, Christianity, and Islam.

In his *Creation of the Sacred*, Walter Burkert adds that "Religion is generally accepted as a system of rank, implying dependence, subordination and submission to unseen superiors."[9] Nicolas Wade asserts that "In marriage and reproductive practices, in enforcing standards of morality in political movements, in generating the bonds of trust essential for commerce, and in warfare, religion continues to play many of its ancient roles as effectively as ever."[10] Richard Rorty goes so far as to conclude, "I think that if the churches gave up the attempt to dictate sexual behavior they would lose a lot of their reason for existence."[11]

But the domination factor or police role of religion is, of course, hardly the whole picture. Religion consoles. Belief in the teachings of a religion serves to answer basic, haunting questions. "Is That All There Is?" is one way to put it, Peggy Lee's 1969 signature song. We are strangers in the world; we

---

9   Walter Burkert, *Creation of the Sacred* (Cambridge, MA: Harvard University Press, 1996), p. 81.

10   Wade, *op. cit.*, p. 17.

11   Richard Rorty and Gianni Vattimo, *The Future of Religion* (New York: Columbia University Press, 2004), p. 79.

no longer belong. We question the growing emptiness and long for sustenance.

Within civilization, religion has played a crucial role in ensuring society's survival. But it is civilization that has made us dependent, worshipful, in need of consolation. Only a "politics" that aims at a reconnection with the earth, with the holistic actuality of nature, will address the spiritual hunger, the need for religion. Wholeness, immediacy, simplicity are not on offer with civilization.

Victoria Hegner and Peter Jan Margry write of *Spiritualizing the City*,[12] and David Lodge and Christopher Hamlin propose ecology and religion for a "post-natural" world in their *Religion and the New Ecology*,[13] both horribly missing the point. Meanwhile, millions seem to have found a spiritual meaning in the iconic, poster-enhanced photograph of the Earth taken by astronauts from beyond the moon. A shot only made possible by a massive industrial foundation, the progressive ruin of the planet. The spiritual certainly lies elsewhere.

---

12 Victoria Hegner and Peter Jan Margry, editors, *Spiritualizing the City* (New York: Routledge, 2017).

13 David M. Lodge and Christopher Hamlin, editors, *Religion and the New Ecology* (Notre Dame, IN: University of Notre Dame Press, 2006).

Things are going to slide, slide in all directions
Won't be nothing
Nothing you can measure anymore.

The Future, Leonard Cohen

DEATH AND THE ZEITGEIST

We are in mass society's Age of Pandemics. At this stage of civilization nothing is stable or secure.

The Age of Pandemics is also the Age of Extinction, as in no longer existing. Death as an existential, ontological matter.

Nursing homes, prisons, meat packing factories—where humans and other animals are warehoused under the sign of Death. Meanwhile, life continues at the extremes of representation, the time of the virtual spectacle. Digital validation is the norm in hypermodernity. What exists is what is on the screen, displayed on the display screen and not elsewhere.

James Poniewozik's "Life. Death. And Something Disturbingly Other," in a recent *New York Times*, finds that as life becomes increasingly digitized and simulated, there is more attention to death, especially in the context of the COVID-19 pandemic. Imagining one's own death comes to the fore.

Death presents itself, but what of life's substance? In the technosphere, our lives seem to have a decreasing plenitude or fullness, less of a possibility for a full life. A foundation is lacking, and so fear of death grows.

The coronavirus pandemic is an overlay on an already reduced, transformed life-world. Its death toll is nothing to discount, but it is just part of the bill coming due for civilization's relentless assault on the natural world.

In terms of our orientation to the death count, French historian Dominque Kalifa on France Culture Radio in May, at the height of the contagion fears, argued that the pandemic, of course serious, has evoked an overreaction bordering on panic. Neither the "Spanish" influenza of 1918–19 nor the AIDS pandemic (170,000 deaths in the U.S. in the 1980s), though more deadly, prompted the 24/7 fixation of 2020.

Comparisons are difficult, however. For example, the U.S. AIDS death count was the measure of a few years, from the early to the mid-1980s, whereas coronavirus fatalities reached 100,000 in only three months. But Kalifa's overall point may be valid.

Pandemic experience may be exhibiting a heightened fear of death, underlined in the techno-culture context. Lauren Collins' article "Missed Calls: Long-Distance Love, Death, and Grief," in a May *New Yorker*, recounts her encounter with her dying father. She was unable to be at his bedside; a phone was pressed to her father's ear. "I listened while Dad gasped for breath, waiting for someone to reclaim the phone. Do you mute something like that if you can?" Deathbed by device, new levels of estrangement. Bound to technology even in death, especially in a pandemic context.

The rapid buildup of pandemic fatalities, the suddenness of mass shooting episodes. Death has a scarier face. In popular culture, the zombie figure looms large—*The Walking Dead*, for instance, the wildly popular television series.

Behind it all is the primary threat, identified by novelist Zia Haider Rahman as "the monster of modernity." Modernity, Enlightenment, Progress have been unmasked, their ideological returns very greatly diminished. But Technology continues to promise everything. One of its variants, transhumanism, even pursues an unhinged claim to deliver triumph over death.

The delusions pile up; what is real becomes elusive. In 2013, Henry Allen penned an op-ed piece in the *Wall Street Journal*, headlined "The Disquiet of Ziggy Zeitgeist." Allen discloses his "sense that reality itself is dwindling." His "disquiet" is further revealed: "For the first time in my 72 years, I have no idea what's going on. I used to be Ziggy Zeitgeist, Harry Hip. I like to think I was especially good on the feeling-tone of the world around me."

Not anymore, and Henry Allen, a former *Washington Post* editor, is not alone in his bewilderment. We are more and more rootless, adrift, less and less grounded. Death is robbed of its meaning, its arrival at the end of a life of connection and consequence. Along this trajectory, as civilization runs its fatal, life-draining course, fear becomes more predominant.

RACISM AND THE SYMBOLIC

Maybe more than ever symbols now loom large. Racist statuary has been a visible target of the Black Lives Matter movement, and one thinks of the original symbolic markers of domination, the monumentalism of the Neolithic era, when domestication had triumphed.

In a sense the symbolism has become democratized. Wearing a mask or not wearing one during the pandemic is a statement for some, symbolizing one's politics. Even something that utilitarian can be weaponized, made to say something outside the reason it exists. Trump's refusal to wear a mask is of clear symbolic intent, for instance.

Race is a symbolic product, constructed to express certain values and results. Its earliest roots lie with domestication, and the need to determine or produce traits in line with the control ethos that is domestication. Self-domestication set in as the domesticators began to assign symbolic, reified judgments to themselves and other humans. "Racial characteristics" and racism resulted.

The displacement and hierarchy of domestication is predicated on the symbolic. Eduardo Kohn (*How Forests Think*, p. 49) put it this way: "Symbolic thought run wild can make us experience 'ourselves' as set apart from everything: our social contexts, the environments in which we live, and even our desires and dreams." Racism is one aspect of this dynamic; it justifies the basic separation as it is lived out in social life.

"All reification is a forgetting" is a well-known formulation by Theodor Adorno and Max Horkheimer. The living entity and its context are forgotten, in order to effect a reduction, e.g., a racist conception or image.

Reification is closely related to objectification; to objectify is to freeze what is alive and various into a thing, an object that receives its alleged symbolic essence or character. Racism in our culture is frozen in time, and has been for millennia. Even time itself is a symbolic product, a mysterious and now dominant reified fact of life. The symbolic is the bedrock short-circuiting of a fuller understanding. No wonder it's so challenging to change our racist beliefs.

The effort to undo racism implies anti-symbolism. It isn't that art, literature, etc. are not employed against racism. Of course they are. But we must not forget the forgetting, or the reason for it. On the eve of domestication's reign, symbolic thought set the stage, prepared the way. The move from immediacy to a world of fixed separations, mediation, and hierarchy required symbolizing, reifying, objectifying modes of thinking.

Flags, monuments, masks—not forgetting race—dominate society's discourse with a weight of oppression. Time to dump this ever-growing load of the inauthentic.

TECHNO-MADNESS

The age of pandemics is doing its share to extend and deepen the technosphere. Cyberspace was to have connected us, seamlessly and instantaneously. But anxiety, depression, and loneliness have only grown, aided by pandemics.

Tom Siegfried's "Self-Destructive Civilizations May Doom Our Search for Alien Intelligence" (*Science News*, July 6, 2020) provides a provocative observation about civilizations and technology. Scientific projections indicate that when a civilization is technologically advanced enough to communicate in space, that civilization is nearing its end. Less likely to sustain signals or anything else.

The implication is obvious: our civilization is being consumed by the very technology by which it measures its "advanced" state. Civilization/technology consumes its host (Earth) and dies.

FASTER! THE AGE OF ACCELERATION

Acceleration is a key fact of human existence today. Time, technology, and modernity are speeding up at an entirely unprecedented rate. These categories or dimensions are becoming parodies of themselves.

Experience, consciousness, our sense of everything is rushed along in a perpetual, hypermediated present—which is not a present. The now has been banished from itself. In the technosphere phase of civilization, our very notion of the present moment has been redefined (mostly) as what flies by on a computer screen. Flying fragments, including us.

Even the perennial Subject-Object question, viz. is there an insuperable alienation between them?, seems to fade in the Age of Acceleration. The subject is ever more insubstantial and disembodied; the object maybe even more so. We are no longer so much surrounded by things as by fleeting virtual images.

Reality seems out of control, in a runaway world. As our daily lives accelerate, less and less happens to us. The power of acceleration, once thought by some to be liberating, is far more widely felt as an enslaving pressure.

Progress "on speed," as it were, heightens the advance of ecological catastrophe. Nature is systematically overburdened, and of course acceleration is its measure.

All of this outraces the ability of thought to come to grips with it. Hartmut Rosa puts it this way: "I fear that we are in danger of running out of claims, hypotheses and theories that are inspiring and challenging for late-modern culture."[1] Though not an acceleration theorist like Rosa, Bruno Latour

---

1   Hartmut Rosa, *Alienation and Acceleration* (Natchitoches, LA: NSU Press, 2010), p. 7.

offers "Why Has Critique Run Out of Steam? From Matters of Fact to Matters of Concern."[2]

But there's no mystery involved. Rosa himself observes that it is "the logic of acceleration that determines the structural and cultural evolution of modern society,"[3] and that "Progress and acceleration were indissolubly linked together from the very beginning."[4] It is clear that growing social and technological complexity, boosted by increasingly interdependent systems and exponentially more powerful computing capacities, constitute the racing reality of mass techno-society.

The reality is more starkly revealed the faster it goes in every sphere. This has not meant, however, that an indictment of the whole is allowed. Such a rejection is perhaps ever more transparently or unavoidably implied, but that perspective is unlikely to be taken seriously. "From the very beginning," in Rosa's words, is an entirely appropriate usage nonetheless. It must be noted in passing that the current nature of this accelerating world is the outcome of the movement of two of the most primary social institutions: division of labor and domestication.

Time has always been a colonizing force. It became a thing, within us, then over us. The emergence and growth of this materiality, time consciousness, corresponds to that of alienation because it is the most primary estrangement. Time is neither neutral nor objective, especially as the depth of the present gives way to the techno-present and our sense of time is re-coded. It leaps forward with all the rest in our epoch, in which even the speed of light, a supposedly

---

2   Bruno Latour, "Why Has Critique Run Out of Steam? From Matters of Fact to Matters of Concern," *Critical Inquiry*, Winter 2005.

3   Hartmut Rosa, *Social Acceleration*, translated by Jonathan Trejo-Mathys (New York: Columbia University Press, 2013), p. 279.

4   *Ibid.*, p. 321.

unassailable limit, has been surpassed. Time cracks the whip and mocks everything that doesn't keep up. It merges with technological existence and in countless ways proclaims that there is nothing outside either of these dimensions.

Time has literally speeded up. We live in a new global timescape that Ben Agger calls "fast capitalism" and the "total administration of time."[5] Along with time compression goes time famine: it feels like we never have enough time; time is running out. Time is getting more and more scarce. Pressure and stress hound us as we struggle, relying on coffee, energy drinks, and other substances to keep up.[6]

The temporal trajectory has become a permanent but impoverished present. As Baudrillard put it, "Time itself, lived time, no longer has time to take place."[7] Domesticated society has long been temporal in nature, now radically so.[8] The new, post-clock age is very decontextualized, but also shows more of the same progress of time's estrangement from the Earth. The present contracts, but is increasingly all there is. History is evaporating; the past becomes somehow incomprehensible. *Posthistoire: Has History Come to an End?* by Lutz Niethammer poses this well enough. Hervé Fischer concludes, "Cybertime, our time, is tragic. It has no past, no future."[9]

---

5   Ben Agger, "Time Robbers, Time Rebels: Limits to Fast Capital" in Robert Hassan and Ronald E. Purser, eds., *24/7: Time and Temporality in the Network Society* (Stanford, CA: Stanford University Press, 2007).

6   The popular literature is of course extensive. For example: Martin Moore-Ede, *The Twenty-Four-Hour Society: Understanding Human Limits in a World That Never Stops* (Reading, MA: Addison-Wesley, 1993); Thomas Hyland Eriksen, *Tyranny of the Moment* (Sterling, VA: Pluto Press, 2001); Brigid Schulte, *Overwhelmed: Work, Love, and Play When No One Has the Time* (New York: Sarah Crichton Books, 2014).

7   Jean Baudrillard, *The Intelligence of Evil or the Lucidity Pact* (New York: Oxford University Press, 2005), p. 27.

8   For historical analysis see my "Beginning of Time, End of Time" in *Elements of Refusal* (Columbia, MO: C.A.L. Press, 1999) and "Time and its Discontents" in *Running on Emptiness* (Los Angeles: Feral House, 2002).

9   Hervé Fischer, *Digital Shock* (Montreal: McGill-Queen's University Press, 2006), p. 50.

There's no time for depth, engagement, reflective action. The symbolic, which started with time, completes itself as the only presence. Fischer again: "Time has become the very matter of reality."[10] If it is true that oppressive time infiltrates and domesticates at a basic level, the struggle against domination cannot overlook it.

Modern technology is precisely what alters our experience of time. The always faster colonization of life by technology commands an ever-fluctuating environment in which the self is destabilized and such dichotomies as online-offline, public-private, and work-leisure are made largely irrelevant. The properties of the physical self are reduced, as galloping technology claims to complete and enhance them. Speed is of the essence; computing power means one thing—how fast it is. 2014's Magazine of the Year award went to a tech and business zine called *Fast Company* (*New York Times*, May 1, 2014).

Staring at screens we become "digital interfaces,"[11] approaching a communicative elsewhere which is nowhere. Through the always developing devices a great indifference to the world is apparent. And why should this be surprising, given how indifferent the world now is to us. A world subdued and rendered uniform, ugly and lifeless by onrushing technology. Enlightenment modernity, its promises unrealized, is now unrecognizable in key ways. Spengler said that modern times have been "stretched and stretched again to the elastic limit at which [they] will bear no more."[12]

The history of modernity is, on one level, a series of innovations in ever-increasing time compression. This mounting

---

10 *Ibid.*, p. 49.

11 Arthur Kroher, *The Will to Technology and the Culture of Nihilism* (Toronto: University of Toronto Press, 2004), p. 175.

12 Oswald Spengler, *The Decline of the West*, Vol. I (New York: Alfred A. Knopf, 1926), p. 19.

technological movement is foundational to the fact that Progress is totalitarian. From urbanization and globalization to the disorienting virtual waves of information, the enclosing pace is relentless.

About a century ago the futurist Marinetti declaimed, "One must persecute, lash, torture all those who sin against speed."[13] Fast-forward to the present: Microsoft Cloud threatens all techno-serfs with the reminder that "The winning edge can boil down to nanoseconds. The Cloud That Helps Win the Race!" (full-page ad, *New York Times*, April 1, 2014). A nanosecond is a billionth of a second. Stock markets around the world now operate on this level. Paul Virilio says that "By accelerating, globalization turns reality inside out like a glove."[14]

Baudrillard stressed that reality comes to an end when real and unreal become indistinguishable. This is the current stage of the catastrophic nature of civilization, which is modernity. And Rosa points to the "acceleration process which is indiscernibly linked to the concept and essence of modernity."[15] Virilio terms this "a culture of desertification,"[16] whose constant uptick guarantees the "liquidation of the world."[17]

All of this operates against mutual and embodied co-presence; this seems to be why (somewhat perversely) all the speeding-up has produced a big increase in sedentariness. Seated before the slight, synthetic glow of the screen as life flies by. As people work faster, because the machines go faster.

---

13 Filippo Tommaso Marinetti, "The New Religion—Morality of Speed" in Hartmut Rosa and William E. Scheuerman, eds., *High-Speed Society* (University Park, PA: Pennsylvania State University Press, 2009), p. 58.

14 Paul Virilio, *The Original Accident* (Cambridge: Polity, 2007), p. 51.

15 Rosa, *Alienation and Acceleration, op. cit.*, p. 8.

16 Paul Virilio, *Negative Horizon* (New York: Continuum, 2008), p. 35.

17 *Ibid.*, p. 52.

"Get.Arts.Fast" is William Grimes' March 21, 2014 *New York Times* offering on abbreviated theater and other performances, shortened to fit the busy schedules of exhausted patrons, as well as their shrinking attention spans. At least accelerating technosphere developments (e.g., Artificial Intelligence, nanotechnology) have so far not managed to keep up with the fantasies of those who actually put their faith in them. Ray Kurzweil's deluded technotopian Singularity dreams get a cinematic comeuppance in the 2014 *Transcendence* movie, by the way.

Somewhere Spinoza wrote that we are immortal here and now, in each instant. And so we continue, in the face of what wants to be overwhelming. Jean-Claude Carrière, asked how he accomplished so many things, responded, "My reply is always the same, and I'm not trying to be funny: 'Because I do them slowly.'"[18] Gifted athletes often remark similarly, on their ability to slow things down.

"All that is solid melts into air," as Marx and Engels characterized the transforming power of industrial capitalism in *The Communist Manifesto*. But what drives the now frantic pulse of transformation was unleashed far earlier. It has indeed gone into overdrive with the Industrial Revolution, but heightening complexity under the sign of domestication is thousands of years older than modern capitalism.

"Don't be evil" is Google's well-known mantra, part of its mission statement. Of course the whole mega-project, of which Google is only the latest tiny part, is the "evil" that is now so virulent. It may appear as the force of destiny, in which case it is time for a new conquest.

---

18 Catherine David, Frédéric Lenoir, Jean-Philippe de Tonnac, eds., *Conversations About the End of Time* (New York: Fromm International, 2000), p. 164.

NOT SO CLOSE ENCOUNTERS: DISTANCED IN THE AGE OF AUTISM

For about a century, scientists have known that all of the galaxies are flying apart from each other, which may be a very apt metaphor for our epoch of social existence. We are hurtling away from each other in a kind of explosion, but our universe, unlike the physical one, is contracting, not expanding.

We are outdistancing—and impoverishing—the natural world as we ride along on a terrible voyage of distancing. The word distance means a stance or standing apart or away. The historical development of civilization and Enlightenment goes on and on, but where are the promised emancipations, where is a basis for community? Instead there is separation, remoteness and their consequences. As philosopher Bruce Wilshire asked, "At the time of the so-called triumph of the West, why do so many people feel so crappy, so lonely, so abandoned?"[1] Health professional John G. McGraw added that "loneliness and other negative states of aloneness can pathologize people,"[2] a phenomenon we see, for example, in growing random violence. "No sense of direction, we Americans./ No place to go," wrote Carolyn Kizer.[3]

Distanced, apart, as we so steadily seem to be, a sense of unreality takes hold. In the functional society of technology, dysfunctional behaviors and conditions abound. Even the incidence of Alzheimer's at ever earlier ages is implicated, as

---

1   Bruce Wilshire, *Fashionable Nihilism* (Albany: State University of New York Press, 2002), p. 101.

2   John G. McGraw, *Intimacy and Isolation* (New York: Rodopi, 2010), p. 420.

3   Carolyn Kizer, "Anniversaries: Claremont Avenue, from 1945," in *Harping On* (Port Townsend, WA: Copper Canyon Press, 1996), p. 59.

a function of loneliness.[4] In 1991 Daniel Goleman addressed "dissociative disorder"—a malady of growing prominence, the third most common psychiatric disorder after anxiety and depression: the feeling that one is an automaton.[5]

The social world is fragmented and there is indeed a worrying trend toward greater fragility and emotional instability. The most severe example is likely autism, for which the prevalence estimates have risen almost exponentially.[6] If the capacity for empathy is a defining feature of human relationships, autistic individuals are more or less cut off from the human experience. Characteristically, they don't like to be touched. Autism literally means "self"-ism and is an inability to form affective, emotional contact with others, living in "a world in which they have been total strangers from the beginning."[7] A remarkable parallel, on a deeper level, with the fact that recent decades have witnessed a striking diminution of contacts with friends and neighbors.[8]

Autism has been described as a marked exaggeration of masculine traits. Heavily affecting males over females, it strongly values systemization, not empathy, and engages with reality minus social interaction. Obsessive, rote behavior (e.g., strict attachment to routines) mimics the high-tech culture, which of course is also a male domain. Temple Grandin, herself autistic and probably the most

---

4   R.S. Wilson et al., "Loneliness and risk of Alzheimer's disease," *Archives of General Psychiatry* 64 (2007), pp. 234–240.

5   Daniel Goleman, "Feeling Unreal? Many Others Feel the Same," *New York Times*, January 8, 1991.

6   Ilona Roth and Payam Rezaie, eds., *Researching the Autism Spectrum: Contemporary Perspectives* (New York: Cambridge University Press, 2011), Introduction, p. 1.

7   Leo Kanner, "Autistic Disturbances of Affective Contact," *The Nervous Child* 2 (1943), p. 249.

8   Aaron Ben-Ze'ev, "Detachment: the unique nature of online romantic relationships," in Yair Amichai-Hamburger, ed., *The Social Net: Human Behavior in Cyberspace* (New York: Oxford University Press, 2005), p. 105.

well-known writer on the subject, has said there would be no Silicon Valley without folks from various parts of the autism spectrum.

These few lines don't even scratch the surface of the reality of the topic. Stuart Murray, who is both parent and healer, observes that "We might care less about causes if we knew exactly what it means to live with autism."[9] As Clara Park put it, "Autism is when your two-year-old looks straight through you to the wall behind."[10] But causes should not be cavalierly dismissed or attributed summarily. Park goes on to refer to how autism "is now almost universally recognized" as the result of genetics, viral infection, and/or unknown biological agents.[11] Very often what is "universally recognized" misses the point completely and is a matter of ideological bias rather than reality.

To confine causation to personal physiology masks what underlies the terrible condition of autism. Environmental degradation and alienation at least as massive are obvious dynamics. Liah Greenfeld's *Mind, Modernity, Madness*[12] argues persuasively that such diseases are culturally caused, that "our inability to cope with [modern] complexity makes us mad."[13] Meanwhile it should come as no surprise that the purveyors of techno-complexity will not see what is right in front of them, even when they provide the evidence. "The Mystery of Autism" is an anonymously penned offering in the December 18, 2014 issue of *MIT Technology Review*, announcing perplexity as to why autism is so prevalent.

---

9   Stuart Murray, *Representing Autism: Culture, Narrative, Fascination* (Liverpool: Liverpool University Press, 2008), p. 211.

10   Clara Clairborne Park, *Exiting Nirvana* (Boston: Little, Brown, 2001), p. 6.

11   *Ibid.*, p. 7.

12   Liah Greenfeld, *Mind, Modernity, Madness: the Impact of Culture on Human Experience* (Cambridge, MA: Harvard University Press, 2013).

13   *Ibid.*, p. 628.

The article's key graphic shows that 1 in 68 in the U.S. are so afflicted, 1 in 55 in Japan, 1 in 52 in South Korea, and so on, vs. much lower incidences in countries that are much less high-tech. The correspondence is not at all perplexing, but is in fact inescapable. Alexander Durig's *Autism and the Crisis of Meaning*[14] puts the connection in unmistakable terms: autism and a wider, societal meaninglessness both follow from postmodern technological society.

Our species is being steadily reconfigured away from humanness toward machine-being, a move of space-time distancing to somehow escape or transcend humanity. Now we can—must?—experience "cyberian apartness," "intimacy at a distance," "detached attachment," and the rest, with the weak and shallow ties inherent in this disembodied zone. Mediated quasi-interaction, essentially non-reciprocal. Atrophy, apathy, cynicism, awaiting robotic companionship, already in place for various Asian elders. A wave of virtualization is taking place, where the Web is increasingly treated and experienced as a real place. Where life retreats, screen-deep, in its "relationships." More and more time online, more depression and boredom.

Raoul Vaneigem referred to the infamy of exchange involved in pushing a coin across the bar. Now we see even that escaping; let's go out to order and pay by touch screen. No human contact at all. Jacques Ellul saw this coming as early as the '80s: "Real meetings will become rarer than ever. We will see one another only by way of machines."[15]

To be distanced is to "lose touch." The first sense to develop, touch is "our most fundamental and complete sense since it allows us to distinguish between an object and

14 Alexander Durig, *Autism and the Crisis of Meaning* (Albany: State University of New York Press, 1996).

15 Jacques Ellul, *The Technological Bluff* (Grand Rapids, MI: William B. Erdmans, 1990), p. 112. Originally, *Le Bluff Technologique* (Paris: Hachette, 1988).

its image copy," notes Alan Kirby.[16] "Haptic" technology has already replicated the sense of touch, in yet another bid to take us away from the directly experienced. Less abstractly, bumper stickers that ask "Have you hugged your child today?" are disappearing, as parents become wary of touching their own children.[17] We are also distanced from what produces the distancing. The unfamiliar, somewhat uncanny devices with screens are hidden from us, not experienced. To be intimate, to be present with this Earth is a challenge; we cannot ignore all that bars that possibility. The challenge has to do with our senses, starting with touch. The world's least tactile people, Americans among them, are the most technologically oriented. Seeing, hearing, tasting, smelling, touching are all on their way to becoming digitalized, computer-designed processes, in flight from the sensual world, from the "animal" life of the body.

The texture of living has joined the exodus of meaning and the motion of the catastrophic nature of modernity presents itself. In 2011 Anders Breivik killed eight people with a bomb, then massacred 69 at a Norwegian youth camp. The novelist Karl Ove Knausgaard, meditating on this horror, concluded that there must be a distance that makes such acts possible. A distance that "has appeared in the midst of our culture. It has appeared among us, and it exists here, now."[18]

"It is difficult to argue against the fact that our twenty-first century has witnessed the emergence of a different breed of human beings under the sign and aegis of the information age," as Paul Jahshan saw it.[19] Twenty years earlier, a character

---

16   Alan Kirby, *Digimodernism* (New York: Continuum, 2009), p. 330.

17   Tiffany Field, *Touch* (Cambridge, MA: The MIT Press, 2001), p. 5.

18   Karl Ove Knausgaard, "The Inexplicable," *The New Yorker*, May 25, 2015, p. 32.

19   Paul Jahshan, *Cybermapping and the Writing of Myth* (New York: Peter Lang, 2007), p. 3.

in Paul Auster's dystopian novel *In the Country of the Last Things* phrased it more darkly: "Life as we know it has ended, and yet no one is able to grasp what has taken its place."[20]

But obviously, none of it happened overnight. Dying horn strains in Schubert's *Winterreise* (Winter Journey) evoke absence and loss, what has passed away, in a world lost to industrialism. Far earlier than any Romantic motif are the distancing mechanisms of civilization's domesticating pulse and instrumental reason. A distant promise that eluded Melville's Ahab and Fitzgerald's Gatsby is now gone, and we are left distanced from ourselves, addicted to diversions and breathing an ever thinner and more troubled air.

Other animals embody a horizon of truth that we, who are not the only animals who mourn, are losing. At the time of a radical eclipse of nature itself—and much else—we need to remember place and particularity. Philosophy is a meditation of place, even though under massive assault by placelessness. Distancing is allowed only as a necessary protection against the force of the very distancing we need to combat. Only that much detachment.

---

20   Paul Auster, *In the Country of Lost Things* (New York: Penguin, 1988), p. 20.

MACHINE PSYCHOLOGY: A DISAPPEARING ACT

Already in the 1960s, Theodor Adorno found the computer to be "the bankruptcy petition of consciousness."[1] Now we are at the threshold of cyborg existence, wherein the self emerges as a shifting matrix of animate and inanimate parts. This is being accomplished insofar as we have reduced ourselves to the machine's level. "The very possibility of subjectivity and the generation of meaning for the future" is at stake, as Gray Kochhar-Lindgren put it.[2] As the techno-culture advances in its fast-forward course, what has become undeveloped is the result. "We are as highly developed in psychopathology as in technology," concluded Jules Henry.[3] Lightning speed of connectivity—and increasing disconnection among people. Considerable regress in substantive communication.

Consciousness, like perception and cognition, is embodied. Except when it's not. Direct, primary experience is severely on the wane. Life has moved to the screen, where all is secondhand experience. Cyberspace is not the realm of the texture, depth and continuity of anyone's life-world. Without the immediacy of human experience we live in a Dead Zone. And we need a tremendous distraction industry or apparatus because the content of our own activity is thus diminished.

Our Age of Distraction[4] privileges external stimuli, especially data, over interior reflection. From understanding to knowledge to information to data, the lowest level in the mental food chain. Undigested information—data—is apt to

---

1   Theodor Adorno, *Negative Dialectics* (New York: Continuum, 1997), p. 206.

2   Gray Kochhar-Lindgren, *TechnoLogics* (Albany, NY: State University of New York Press, 2005), p. 1.

3   Quoted in David Levin, *Pathologies of the Modern Self* (New York: New York University Press, 1987), p. 480.

4   Joseph R. Urgo, *In the Age of Distraction* (Jackson, MS: University Press of Mississippi, 2000), e.g., p. 19.

become disinformation, confusion, deception. Discernment, attention span, etc. are so many casualties of the erosion of deeply felt experience.

Thought is not data-processing, however much the metaphor of an essentially machine-like process tends to creep into our consciousness. We "scan" this, "process" that, in common usage. Faster and faster speeds online "input," and we are increasingly impatient with the slightest delay. But deep experiences require more than fractions of seconds. More and more reliance on cyberspace means we know less and less about the world available to us directly.

Technological systems are configured to be mind-deadening. The ultimate danger is that people will become addicted to technologies that in effect ban their capacities to think, that take away a basic sense of identity and reality. For some time now, we've been immersed in a universal culture of redirection, diversion and infotainment, a mass-based content level of cognitive distraction. Unsurprisingly, boredom and anxiety are hallmarks of this restless ethos. Attention is at a premium, sought and manipulated like data.[5] Interacting face-to-face becomes rarer. A barren, synthetic reality challenges us to be capable of thinking for ourselves, because without reflection, hope for a different world dims.

Prominent among those who embrace the technoverse is Donna Haraway, who claims that machines do not dominate or threaten us: the machine is us. Too late to check the invasiveness of technology anyway: "Prothesis becomes a fundamental category for understanding our most intimate selves."[6] Susan Griffin counters this beautifully: "We know ourselves to be made from this earth. We know this earth

5   See Adam Gazzaley and Larry D. Rosen, *The Distracted Mind* (Cambridge, MA: The MIT Press, 2016) and Steve Talbott, *Devices of the Soul: Battling for Our Selves in an Age of Machines* (Sebastopol, CA: O'Reilly Media, 2007).

6   Donna Haraway, *Simians, Cyborgs and Women* (New York: Routledge, 1991), p. 249, n.7.

is made from our bodies. For we see ourselves. And we <u>are</u> nature."[7] The tragedy is that only machines thrive in the global technoculture. Some people grow to be like machines, à la Haraway.

All this is new and, in a sense, not new. Kochhar-Lindgren proposes that this is "when philosophy becomes what it has always latently been: cybernetics."[8] In other words, philosophy's control and alienation aspects have been lurking for a long time. Since its origin in early civilization, writing has been an independent object, abstracted from reality. It virtualizes and delocalizes memory. The text stands over us in a sense, like time. The decline from speech to writing to the increasing digitalization of human consciousness exacts a price.

Society becomes an unhealthy expression of the Internet. Michael Heim calls this "a unifying network of human presence."[9] Of course, we are not present to each other online, and there is no substitute for actual place. Marc Andreessen had it right in 2011: "Software is eating the world."[10] Phenomenologists speak of being-in-a-world, meaning that it is from the world that we come to understand not only it, but ourselves. The technosphere is immersive and invasive. It patterns or structures our lives. Our sense of place, presence, and self is being remade in the image of the Machine.

---

7   Susan Griffin, *Woman and Nature: The Roaring Inside Her* (New York: Harper & Row, 1978), p. 226.

8   Kochhar-Lindgren, *op. cit.*, p. 5.

9   Michael Heim, "The Metaphysics of Virtual Reality," in Sandra K. Kelsel and Judith Paris Roth, editors, *Virtual Reality: Theory, Practice, and Promise* (Westport, CT: Meckler Publishing, 1991), p. 34.

10 Marc Andreessen, "Why Software Is Eating the World" (*Wall Street Journal*, August 20, 2011).

Champions of cyborg existence applaud the "seamless integration and overall transformation" of human and machine.[11] Smartphones are not zombifying agents of distraction, but "mindware upgrades."[12] Much is disappearing as the sentient and tangible are de-realized. The sense of community, for instance, shrinks in proportion to the extension of global online culture. An exception to disappearance is social media; what is there is eternal. Kate Eichorn's *The End of Forgetting*[13] discusses how one is hard pressed to break away from the past after those endless and permanent self-surveillance posts to Facebook, Instagram, etc.

Technology has become the organizing principle of our lives. It is also clear that it is civilization's cardinal value, the chief reason that there is just one, global civilization now. Thus it is a huge challenge to be outside it, or to even imagine such a change. And yet individuals are far from happy within the force field of the ever-present technoverse. Its promises and claims are threadbare and false. Everyone knows that, at least on a visceral level. We know that the totality is built on lies and is failing. Consciousness may yet become better armed, more able to take advantage of Walter Benjamin's dictum: "The smallest cell of visualized reality outweighs the rest of the world."[14]

---

11 Andy Clark, *Natural-Born Cyborgs* (New York: Oxford University Press, 2003), p. 34.

12 *Ibid.*, p. 10.

13 Kate Eichhorn, *The End of Forgetting* (Cambridge, MA: Harvard University Press, 2019).

14 Quoted in Adorno, *op. cit.*, p. 303.

ABANDON THE DEATH SHIP

The globalized and globalizing world is a restless, synchronous world of traffic. Under the sign of a now unitary civilization, there is a "single global domestication machine."[1] Its inner logic was well expressed by Jean-François Lyotard: "Development is produced by accelerating and extending itself according to its internal dynamic alone."[2] The totalizing nature of globalization means that all of its aspects are interdependent, an integrated whole that integrates, amalgamates the rest into its web. Like cancer, growth is of the essence, and it has advanced so far that it almost seems bizarre to ask that it be justified.

But present and future require just such a questioning. "Catastrophe is everywhere and ever-present," according to Braun and Wakefield, who see "a civilization already in ruins... we are already living in a post-apocalyptic condition."[3]

Of course, globalization has meant the global extension of what already existed: "the taming for civilatory purposes," in Peter Sloterdijk's words.[4] Technological superiority was the central entitling dynamic over indigenous peoples. Cultures were subdued and deformed, but each managed to hang on to some authentic substance.

Metastasizing toward modernity, milestones were reached and built upon. Gutenberg's printing press in the 1400s paved the way for a more uniform political order; the modern

---

1   John Zerzan, "The Modern Anti-World," in *Twilight of the Machines* (Port Townsend, WA: Feral House, 2008), p. 77.

2   Jean-François Lyotard, *L'inhumaine* (Paris: Éditions Galilée, 1984), p. 14.

3   Bruce Braun and Stephanie Wakefield, "Inhabiting the Postapocalyptic City," *Society & Space*, February 11, 2014.

4   Peter Sloterdijk, *In the World Interior of Capital* (New York: Polity, 2013), p. 165.

insurance system began to emerge in the early 1600s. A literal globalization was coming on with some rapidity.

With our new millennium an "anti-globalization" movement went forth, from 1999 to 2001. But it almost exclusively aimed at dysfunctions of the world system, focusing on the excesses of transnational corporations. "Alter-globalization" became its more accurate, leftist moniker, the desire for a nicer, kinder globalized world. Sloterdijk, however, is accurate in noting that there were some who "openly stated their belief that it would have been better if humans had never reached the global stage."[5]

It is obvious that technology is a critical part of the international system, the component that underlies all of it. The novelist Don DeLillo's "In the Ruins of the Future" defines a superpower in terms of technology. The fabric of the modern system is the Internet, our material and immaterial culture, its embedded principle and its means of governance.

Henry Kissinger et al. wrote recently of the "unstoppable" Artificial Intelligence revolution. A form of digital utopianism is on offer, in which truth and reality themselves are being redefined.[6] Digitality is the hallmark of globalization now—or is it more accurate to say that the hallmark of digitality is globalization? Both are fundamental to the universalization that is civilization. Is technology not at least as decisive as capitalism?

Yuval Noah Harari sees Dataism as a new religious epoch, resting upon Big Data and its attendant algorithms.[7] An always accelerating ethos wherein everything moves toward the reified goal of being reduced to data.

5   *Ibid.*, p. 163.

6   Henry A. Kissinger, Eric Schmidt, and Daniel Huttenlocher, "The Metamorphosis," *The Atlantic*, August 2019.

7   Yuval Noah Hariri, *Homo Deus: A Brief History of Tomorrow* (New York: Barnes & Noble, 2016), e.g., p. 373.

Modernism rides the technological wave, which sets the pace and defines what fits. Once in India I asked how people manage to accept the severe air pollution. A seemingly common answer had to do with allowed aspirations; namely, many aspire to drive an air-conditioned car to an air-conditioned office, returning to an A/C apartment. Such pollution is a global problem, with cities, the key sites of civilization, increasingly becoming disaster zones in general. In the late second decade of the millennium, for example, the air in London was more toxic than that of notoriously polluted Beijing. In that decade IBM's well-known mantra was "Let's Build a Smarter Planet," shifting to "Smart Loves Problems" at the end of the decade, as if to acknowledge, however obliquely, that "problems" remain so very much with us.

There are hundreds of other environmental problems comparable to air pollution, which I needn't bother to list. They are all too apparent and pressing. Worldwide, these problems disproportionately affect people living in poverty, especially people of color. It is a stark fact that the faster technology rushes forward, the worse the overall condition becomes for Earth's biosphere and each of its species, including us.

The techno-world is a non-place, existing everywhere and nowhere. Homogenized, lacking texture and an elsewhere. In 2000 Pico Iyer's *The Global Soul*[8] described an increasingly homeless, rootless world in which airports, malls, apartments, schools, etc. are identical; people everywhere dress the same, and watch the same television. Universal machine consciousness in unlivable mega-cities. The emoji era, when humanity flows away from humans and other species and toward Technology, the source of what now passes for substance and meaning.

---

8   Pico Iyer, *The Global Soul: Jet Lag, Shopping Malls, and the Search for Home* (New York: Knopf, 2000).

The hallmarks of hypermodern alienation are isolation and loneliness. When community has been dissolved by technological mass society, social existence is fundamentally undermined. The horror of daily mass shootings is a constant in the U.S. The obvious question as to what kind of society spawns these bloodbaths is absolutely prohibited in media-based reporting, one example of the blindness of pathological civilization.

Utter dehumanization means far fewer social ties and solidarity. *The Connection Gap: Why Americans Feel So Alone* writes Laura Pappano[9]; *Full Catastrophe Living* is Jon Kabat-Zinn's offering from the Stress Reduction Clinic of the University of Massachusetts Medical Center.[10]

The technosphere renders us deskilled and dumber. GPS technology, for instance, means that we lose the ability to read a map or have a sense of direction. Our sensibilities, imaginations, depth of thinking, attention spans wane; the literature is enormous, the studies endless. Suicide rates rise, life expectancy decreases, opioid deaths are rampant. Sedentary life means chronic ill health, including diabetes, cancer, heart disease, as every measure of physical activity from youth sports to golf, registers decline. World health is ravaged as the virus of global development pushes forward: Zika, Ebola, dengue fever, West Nile, etc.

How about a real anti-globalization/anti-civilization movement?

---

9   Laura Pappano, *The Connection Gap: Why Americans Feel So Alone* (New Brunswick, NJ: Rutgers University Press, 2001).

10   Jon Kabat-Zinn, *Full Catastrophe Living* (New York: Bantam Dell, 2005).

CAN'T GO BACK

Somewhere, in passing, Herbert Marcuse wondered how the past might be redeemed. A tall, possibly unimaginable order. The past may now seem even further from any conceivable redemption. Are we not already somehow "past"? Is there not a feeling of extinction in the air, including our own extinction?

There's religion and psychoanalysis to repair an individual's past, but society goes wanting. Enlightenment confidence, with its agenda of mapping, measuring, improving, projected a triumphant future. But as Horkheimer and Adorno point out, "the fully enlightened earth radiates disaster triumphant."[1]

Soren Kierkegaard rejected Hegel's projected "cunning of reason"[2] path to the victory of Spirit. The rational world has shown itself to be a path to disaster, not a voyage of healing, as Kierkegaard foresaw.[3] Much later, but over a half-century ago, Adorno firmly grasped the coming physical catastrophe: "There is a universal feeling, a universal fear that our progress in controlling nature may increasingly help to weave the very catastrophe it is supposed to protect us from."[4]

Studies at every level detailing worsening environmental doom come forth with almost daily frequency. Thousands of specialists routinely report that their earlier studies, however alarming, were not alarming enough. Our condition is little more than that: sped-up, weightless particles of data, set in motion by the very forces that are putting this last civilization into its terminal stage.

---

1   Theodor Adorno and Max Horkheimer, *Dialectic of Enlightenment* (New York: Continuum, 1979), p. 3.

2   G.W.F. Hegel, *Lectures on the Philosophy of World History* (Cambridge, UK: Cambridge University Press, 1975), p. 89.

3   Naomi Lebowitz, *Kierkegaard: A Life of Allegory* (Baton Rouge, LA: Louisiana State University Press, 1985), p. 3.

4   Theodor Adorno, *Negative Dialectics* (New York: Continuum, 1997), p. 67.

Can reality be repaired, retrieved? *Historia Ludens: The Playing Historian* (2020)[5] is a gruesome but not atypical approach that serves to put us even further asunder. It actually poses an answer for historiography with video games as its model. How far down the rabbit hole of digital "reality" have we gone with gaming as key category in historical scholarship, when history is understood as a form of such playing? All part of the Machine; even the past has become computable.

We've been on this road for a while. The path of fragmented, synthetic "experience," away from what is life-centered, re-embodied. Looking for an understanding of the role of technology in society is ultimately to grasp what characterizes modern life in late civilization. Movies like *Her* and *Transcendence*, Dave Eggers' novel *The Circle*, and the like expose what is happening, but the force field of the technosphere does not seem diminished.

Data is the basic currency of our age. Its embrace is very hard to avoid, and its ascendance is the mirror image of a place of isolation and loneliness. Separation is its nature. A datum is not a subject, has no living history. Data is not where the past can be understood or redeemed. Efforts to regulate or reform the Data World obscure its nature and basic destination.

An ever colder world conditions us to come to turns with the simulations it provides. The succor of fake connection is better than no connection at all. ELIZA was a 1960s computer program that aped counseling dialogue by means of a simple feedback approach. For example, if one complained about a friend, ELIZA might respond, "Tell me about your friend," or turn a statement into a question. It was striking that even when a subject was aware that ELIZA was not human but rather crude, computer-generated responses, the subject behaved as if it were human. Complete with appropriately

---

5   Alexander von Lunen, et al., *Historia Ludens: The Playing Historian* (New York: Routledge, 2020).

emotional involvement. Many decades ago, we were vulnerable to overlooking the human/machine divide.

In the fully digital era the cyborg—merger of human and non-human—comes of age. "'Cyborg Art' as a Critical Sphere of Inquiry into Increasing Corporeal Human-Technology Merger" by Elizabeth Borst[6] explores the profusion of cyborg art; and its title speaks volumes. Zeynep Gunduz' "Digital Dance" continues in this vein. Because of the disembodiment caused by the intrusion of digital technologies, Gunduz argues, there must be a "re-negotiation of bodily boundaries." How to manage the incorporation "of technologies into the domain of the body"?[7]

Domestication of Media and Technology (2006)[8] is about the enactment of the new technology, how we are made to feel 'at home' with it. Of course, it is possible to ignore the tech mention altogether, even in terms of basic social and political terrains. Brian Caterino and Phillip Hansen offer *Critical Theory, Democracy, and the Challenge of Neoliberalism* (2019),[9] a very Habermasian treatment containing not a word about the pervasive technological context.

More outlandish is *The Quiet Avant-Garde* (2019)[10] by Danila Cannamela, which counsels that it's time to bring back Futurism. The Futurists worshipped the machine and

6   Elizabeth Borst, "'Cyborg Art' as a Critical Sphere of Inquiry into Increasing Corporeal Human-Technology Merger," in Daniel Riha, ed., *Frontiers of Cyberspace* (New York: Rodopi, 2012).

7   Zeynep Gündüz, "Digital Dance: Encounters between Media Technologies and the Dancing Body," in Riha, *op. cit.*

8   Thomas Berker, Maren Hartmann, Yves Punie and Katie Ward, *Domestication of Media and Technology* (New York: Open University Press, 2006).

9   Brian Caterino and Phillip Hansen, *Critical Theory, Democracy, and the Challenge of Neoliberalism* (Toronto: University of Toronto Press, 2019).

10   Danila Cannamela, *The Quiet Avant-Garde* (Toronto: University of Toronto Press, 2019).

speed, and soon became Fascists, but hey, now is the time "in which humans and nonhumans intermesh."[11]

Management CEO Ben Waber's *People Analytics* (2013)[12] tells us that the "future consists of connection, collaboration, and data.... It's a future where age-old practices of relationship building and trust are married with the new age of data gathering that the world of sensors and data streams has brought forth."[13]

But the collapse of community means that digital technologies provide the illusion of connection—connection/communication without a world. "Online community" is a hoax. The claim that regeneration of the social must involve a tech-based recovery of community through networks only masks the fact that such "communities" are, at base, only products and commodities, not human bonds.

It can be argued that the formal aspects of art provide a truer account of history than does historical documentation. As Adorno put it, "The state of truth in works [of art] corresponds to the state of truth in history."[14]

For instance, it was possible to write authentic popular music in the 19th century, but the genre has long since decayed, reflecting a progressively impoverished overall culture. The repetition and predictability of contemporary popular music obey an industrial formula; the standardized reactions it evokes condition the listener to the standardized patterns in society.[15] The harmonious resolution of pop music cannot be reconciled with the governing totality,

---

11 *Ibid.*, p. 7.

12 Ben Waber, *People Analytics* (Saddle River, NJ: ET Press, 2013).

13 *Ibid.*, p. 201.

14 Theodor Adorno, *Night Music: Essays on Music 1928–1962* (New York: Seagull Books, 2009), p. 86.

15 See my "Tonality and the Totality" in *Future Primitive Revisited* (Port Townsend, WA: Feral House, 2013), pp. 45–63.

which permits no such resolution. It must be judged objectively untrue.

Adorno wrote of "late" music in which music critiques itself, takes leave of itself. Beethoven's late quartets, in particular, come to admit how broken, how homeless, is our reality, and in so doing, come to a sort of rescue mission.[16] Dissonance and atonality describe a fractured world. The limits of music are faced; the limits of our condition are no longer glossed over in a false reconciliation.

With music, as with every other cultural vehicle, "Totality is not available to representation, any more than it is accessible in the form of some ultimate truth," declared Frederic Jameson.[17] What hasn't the Internet colonized? Maori poet Robert Sullivan replies in two lines:

> So many sunsets
> Facebook and Instagram couldn't contain them.[18]

Writing in the November 2019 *Artforum*, Erika Balsom reviewed Brett Story's film "The Hottest August." She summed up its message thusly: "At a time when a sense of shared reality lies in ruins, it imagines life as a collective endeavor taking place in a world held in common."[19] Held in common, but not owned.

Samuel Burkeen's 12/13/2019 letter in the *Wall Street Journal* hits the mark: "Civilization is doing more than marching backward. It is in a gradually accelerating self-annihilation mode."[20] All that remains is the obvious conclusion: we must end it. This is the work of redeeming the past.

---

16   Adorno, *Night Music, op. cit.*, "Beethoven's Late Style," pp. 11–18.

17   Frederic Jameson, *Jameson on Jameson*, Ian Buchanan, ed. (Durham, NC: Duke University Press, 2007), p. 35.

18   Robert Sullivan, "Hello Great North Road," *Poetry*, July/August 2016.

19   Erika Balsom, "The Hottest August," *Artforum*, November 2019.

20   Samuel Burkeen, "Civilization's Backward March Not Led by Measles," *Wall Street Journal*, December 13, 2019.

OUT WITH THE NEW

"Have a seat, take a load off." Restful, no doubt healthful, what could be wrong with sitting in a chair? In 2016 the American Heart Association warned that in order to cut down on diabetes and cardiovascular disease we need to sit less. Since then various articles and studies have underlined this finding. Sitting has become the new smoking, it seems.

Chairs certainly were not always so commonplace. Even after the Middle Ages, chairs with arms and backs were reserved for the elites. The common people had benches, stools, barrels and the like. In this vein, there was white bread for the rich and well-placed, and coarse dark bread for the masses.

But what was allotted was not always what was less healthful, and not just in terms of bread. To take this much further back in time, consider "Sitting, Squatting, and the Evolutionary Biology of Human Activity" (*Proceedings of the National Academy of Sciences of the United States*, March 2020), which looks at hunter-gatherer modes of resting, finding them quite superior to sitting in chairs. The study reports low levels of muscle metabolism associated with chair sitting, versus higher levels of muscle activity of the more common squatting and kneeling in non-industrial hunter-gatherer societies.

As a group of us in south India ate with our hands from food on a large banana leaf, an Indian friend told me that "eating from a plate with cutlery is like making love with your clothes on." That simile had never occurred to me, a modern Westerner, that direct, intimate approach to food.

The fork did not come into general use throughout Europe until the 18th century. Norbert Elias' *The History of Manners* (*The Civilizing Process*, vol. 1) discusses changing table manners and changing attitudes about sex, nudity, and hygiene from 1400 to 1800. The historical direction is toward repression, with etiquette and an increasing sense of shame concerning the body and bodily functions.

During the coronavirus pandemic one of the most acute sources of contemporary anxiety in the U.S. was fear of running out of toilet paper, a major crisis in itself for many. Of course, most of Earth's people employ other means. If everyone used toilet paper, global deforestation would be complete.

Indonesia is the second biggest plastics polluter, after China. But now there are new projects, new approaches that are anything but new. Seaweed for containers: it grows freely, without deforestation or pesticides. Also cassava fiber and sugar cane alternatives to plastic, for bags, containers, etc. See Mongabay, March 25, 2020.

Some might give pride of place to plastic as the most important, most consequential invention of all time. The perennial winner in that category, however, is the wheel. First introduced for pottery, not transport, in a matter of three centuries the wheel was enabling horse-drawn chariots in the Caucasus region. This was about 3500 B.C., already in the Bronze Age of metal alloys. According to Elena Kuzmina's *Prehistory of the Silk Road*, metal in the form of weaponry was needed to protect mobile livestock property. Wheeled transport was a further element of domestication, enabling an explosive expansion of exchange. Kuzmina concluded that it determined the course of civilization (page 35).

The appearance of permanent, stronger dwellings also had a lot to do with protection of property. Settled life itself did not occur until virtually the end of prehistory, the coming end of mobile hunter-gatherer freedom. Domestication and sedentism, with almost no exceptions, went hand in hand.

Jerry Moore (*The Prehistory of Home*) claims that [other] animals have shelters, while only humans have homes.

I'm a bit in the dark as to what he means. A further, and clearer, statement may be relevant. Only human animals have ruined the global habitat. And earthquakes don't often kill people; heavy structures falling on them do.

Architecture without Architects by Bernard Rudofsy, and Jean Dethier's *Art of Earth Architecture* are useful resources. Much more so is the wealth of indigenous experience to learn from—for example, the ancestral tule reed and redwood bark dwellings of Ohlone people of the San Francisco Bay area.

So many wonderful, inventive steps forward and here we are, being promised the next gifts of domestication. Further isolating, enfeebling, dependency-deepening productions as collapse proceeds on all fronts! We need a Return or it's game over.

This possibly last civilization has shown us, in its convulsions, the path to ever-worsening anti-life conclusions. It has the deepest roots but it is now, arguably, all the easier to see them. Civilization, as it has globalized, leaves no room and no way out but for its end.

LIFE AS CIVILIZATION BEGINS TO CRUMBLE

I've long had the feeling that what we try to say via politics and theory could be more fully expressed in the language of health or non-health. We can grasp much more in that key, on that all-important level. Our lives as lived are the basis of the nature of society. The reigning loneliness, for example, deep and uniquely modern, tells us more than all the jargon of social theory. Almost entirely omitted from philosophy down through the years, the dis-ease, the suffering, is the indispensable Ground Zero of understanding. The body doesn't lie.

The sorrow that is civilization takes many forms, reaches us everywhere in countless ways. Freud's *Civilization and Its Discontents* located the primary source of unease, of neurosis, in domestication; its fullness is a broad vista of allied pain and loss. The loss of an original immediacy and freedom sets the worsening course for a dark present. Not forgetting that degenerative and infectious maladies originate with domestication.

R.D. Laing once wrote that only by "the most outrageous violation of ourselves have we achieved our capacity to live in relative adjustment to a civilization driven to its own destruction" (*The Politics of Experience*, 1967, p. 64). But Freud's point is that the adjustment is indeed only relative. It comes with an enormous price tag. Russell Jacoby hit the mark: "The modern individual is in the process of disintegration" (*Social Amnesia*, 1975, p. 177).

The steep decline in mental health across mass society is a stark reality. "Why is America So Depressed?" ponders Lee Siegel (*New York Times*, January 3, 2020), noting a 36 percent increase in depression and anxiety from 2016 to 2017 and the fact that nearly 20 percent of Americans have an anxiety disorder. "Anxiety is on the rise in all age groups," according to Maria Russo ("9 Books to Calm an Anxious Toddler," *New York Times*, January 18, 2020).

The mental health crisis is especially acute among the young. The rise in school shootings over the past two decades is one factor among many, leading to higher levels of antidepressant prescription rates. According to Frederic Jameson, anxiety is the expression of "an underlying nightmare state of the world" ("Regarding Postmodernism," *Social Text* No. 21, 1989).

Most Americans have one or more chronic health problems, from emotional disorders to high blood pressure, Lyme disease, arthritis, intestinal upsets, insomnia, etc. One hundred million suffer from chronic pain, a shockingly high number that is rising. "This Won't Hurt" by Sophie Elmhirst (*1843*, September 28, 2019) reveals that fully one-fifth of the world's population endures pain as a regular condition. A significant percentage of television advertising addresses chronic health afflictions, a somewhat new development. Joann Kaufman's "Think You're Seeing More Drug Ads on TV? You Are, and Here's Why" (*New York Times*, December 24, 2017) explores this phenomenon.

Chronic medical conditions shorten life. Longevity is no longer rising. "Gains in Life Expectancy in the U.S. May Be Slipping," wrote Abby Abrams (*Time*, July 24, 2014). "Life Expectancy for Women is Slipping Overall," noted Marissa Cevallos (*Los Angeles Times*, June 15, 2011). Stephen Castle reported on the British situation, a first in its modern history: "Shortchanged: Why British Life Expectancy Has Stalled" (*New York Times*, August 30, 2019).

Suicide has rightfully also garnered much attention in recent years, in startling terms. Patricia Cohen's "Midlife Suicide Rises, Puzzling Researchers" (*New York Times*, February 19, 2008) recounted CDC findings that the suicide rate among middle-aged Americans went up 20 percent in the previous five years. Another Center for Disease Control study led Brianna Abbot to conclude "Youth Suicide Rate Increased 56% in Decade" (*Wall Street Journal*, October 17, 2009). Clusters of teen suicides have become fairly common, and suicide has become the second leading cause of death among those ages 15 to 24.

The darkness of late civilization casts a pall on all life. All the crises of social existence come together at the level of personal health and well-being—that is, the lack of it. *Deaths of Despair and the Future of Capitalism* by Anne Case and Angus Deaton (2020) expresses a lot by its title alone, and focuses on the threatened condition of middle-aged men. James Tyner's *Dead Labor: Toward a Political Economy of Premature Death* (2019) discloses the profits that accrue from early death in society as life is progressively devalued.

Loneliness has never before been seen as a sign of the times, a social fact, a public health menace. Lynn Darling (*AARP Magazine*, January 2020) asks "Is There a Cure for Loneliness?" The cover of the *Times Literary Supplement* for May 29, 2020 features Adam Foulds' review of two books on the subject, under the heading "The Modern Problem of Loneliness." It is an increasingly lonely, empty world. Mary South's "You Will Never Be Forgotten: Stories" (*New York Times Book Review*, April 26, 2020) discusses the toll on women in a technology-dominated society. The review stresses her verdict on online interactions as "wrecking our sense of community, our relationships, and even our bodies."

"How Our Phones Became our Whole Lives in Just 10 Years" (NBC, January 1, 2020). Technology has invaded, enveloped our lives to an unprecedented degree. A nursing home resident in western Michigan, in pain and dying, cried out to her Amazon Echo for relief: "Alexa, help me... I am in pain!" LouAnn Dagen, 66, died the next day. It has become this sad and grotesque, begging a machine for relief.

In my 2018 essay "Pandemic," I discussed the 20-year mass shooting phenomenon. Since then, of course, there has been an actual pandemic, which, in this hemisphere, really began in 1518 when smallpox arrived from Spain. Soon the disease helped the Spanish conquer the Aztecs and Incas, and greatly aided the genocide of indigenous people in North America.

The coronavirus pandemic has exacerbated existing conditions. *SFGate* (May 27, 2020) reported that "A third of Americans are showing signs of clinical anxiety or depression." Isolation, including dying alone, was underlined by shelter-at-home directives. There was a jump in drug overdoses due to increased social isolation, as well as in obesity, which in turn aggravates coronavirus effects. *Science News* (May 24, 2020) discussed "How Coronavirus Stress May Scramble Our Brains." Sleep disorders, too, become magnified by pandemic pressures. But such problems only echoed the pre-virus reality; they were not created by it.

Background to pandemics is the enormity of the global environmental crisis/collapse. Abrahm Lustgarden gets right to the point with his "How Climate Change is Contributing to Skyrocketing Rates of Infectious Disease" (*Propublica*, May 7, 2020). A catastrophic loss in biodiversity, systematic destruction of wildland, and global overheating have allowed disease to explode. The connection between the spheres of climate and pandemics—and health overall—cannot be missed.

The overlap is obvious: health and environment are inextricably tied together. It's not only pandemic that heightens stress. For instance, "Hotter Weather Linked to Increased Stress and Other Psychological Problems," reported Susan Perry (*MinnPost*, March 31, 2020). Allergy seasons worsen as record pollen counts accompany global warming.

And a pandemic more serious than coronavirus is that of planet-wide air pollution. Ninety percent of urban inhabitants, according to the World Health Organization, are exposed to particulate air pollution that exceeds safety standards. There were ten million premature deaths in 2019 (*Fast Company*, March 2, 2020). Outdoor air pollution cuts an average of three years from human lifespans (*The Guardian*, March 3, 2020), while indoor air quality is also polluted. Office spaces in general are significantly contaminated; "You're Polluting Your Office Just by Existing," offered *Futurity* (October 4, 2019). Amadeo D'Angiulli investigated "How Urban Air Pollution is Linked to Kids' Cognitive Decline" and susceptibility to dementia later in life (*Spotlight on Poverty and Opportunity*, September 29, 2019).

Mental health, chronic disease, life expectancy, suicide rates, loneliness and despair, the climate crisis—all related aspects of the totality, a sad civilizational dusk. Shadows fall across our world and lengthen; the massive weight of estrangement is palpable. The direction it's all headed gets ever clearer. In that realization lie the grounds for hope, the understanding of how very much must be overturned.

PHILOSOPHY/ANTI-PHILOSOPHY

There has been a radical energy afoot that demands expression. Black Lives Matter and related efforts against oppression couldn't be more needed.

At various historical moments of contestation one can hear the cry, "Enough theory, this is the time for action." But theory or analysis has its job to do as well. Theodor Adorno went so far as to assert that theory IS action. That is, theory is required for practice to go forward.

I hope that these explorations may in some small way help us push ahead at the necessary depth.

THE PUZZLE OF SYMBOLIC THOUGHT

The achievements of symbolic culture—what can be expressed through art or language, for example—have been powerful consolations. But they have not been able to save us. "eARTh" is an evocative coinage, but reflects wishful thinking. Art is our compensation for having moved away from, and against, the earth.

In Plato's allegory, we are mesmerized by shadow-images on the wall of a cave, projected by a fire behind us. Instead of turning around to experience reality outside the cave, we go on staring at images, representations, screens.

Regarding the symbol and representation, Jacques Lacan put it simply, "Its condition is that of being not what it represents."[1] Think of René Magritte's painting of a pipe, captioned "This is not a pipe."

Symbols communicate by referring to other symbols, but if we knew only representations we would not be able to evaluate their validity or success *as* representations. Nonetheless, reification is basic to symbolic communication: that which was living becomes a thing. The world presents itself to us, and we re-present it.

Why? How did this arise? All we really know is that symbolic thought is now wholly taken for granted. Leibniz concluded, referring mainly to mathematics, that human knowledge cannot avoid the use of symbols.[2] But there are no numbers in the world, and human knowledge (e.g., manufacture and use of stone tools) predates known symbols by two million years.

---

1   Anika Lemaire, *Jacques Lacan* (Boston: Routledge & Kegan Paul, 1977), p. 55.

2   Massimo Ferrari and I.O. Stamatesen, eds., *Symbol and Physical Knowledge* (New York: Springer, 2002), Introduction, p. 5.

Sibel Barut Kusimba contends that early "sharing relationships... were the first symbolically constituted" ones.[3] Yet only humans have created systems of representation, while many other species have non-symbolic sharing relationships. Others assert, with no supporting evidence, that human consciousness doesn't exist without representation.[4] Michael Franz Basch claims that a central function of the human brain is to turn experience into symbols.[5]

The word comes from the Greek *symbolon*, to bring together. There is a resemblance to the Latin *religare*, re-tie together, the root word of religion. Out of a need, or a loss, both symbol and religion seek to make up for something that went missing. These remedies being symbolic, they cannot really heal.

In Hegel's *Logic*, the very structures of reality are made up of autonomously generated representations. How has all this representing advanced our understanding? Didier Debaise points to part of its failure in his *Nature as Event*: the symbolic enterprise has not helped "to deepen or develop our experience of nature." Rather, it has only served to "obscure its meaning."[6]

It is commonly said that communication involves the transmission and/or reception of symbols. Stephen J. Newton goes a step further: "There is no communication without symbols."[7] Of course, there was face-to-face communication

---

3   Sibel Barut Kusimba, *African Foragers* (Lanham, MD: Altamira Press, 2003), p. 91.

4   For example, Eric Gans, *The End of Culture* (Berkeley: University of California Press, 1985), p. 25.

5   Michael Franz Basch, "Psychoanalysis and Communication Science," *The Annual of Psychoanalysis*, No. 4, 1976.

6   Didier Debaise, *Nature as Event* (Durham, NC: Duke University Press, 2017), p. 1.

7   Stephen J. Newton, *Painting, Psychoanalysis, and Spirituality* (New York: Cambridge University Press, 2001), p. 75.

for thousands of generations before the earliest evidence of symbolizing. At some unknown time, speech began to emerge, very likely accompanying the gradual emergence of social complexity. Symbolic ritual seems fairly recent, pointing in the direction of other symbolic storage systems, like written records. Technologies of symbolic communication across distances followed. The modern age brought the printing press, newspapers, telegraph, telephone, computers.

Communication predates even our existence as human species; symbolic communication is quite recent. Not only is the symbolic a reified form of communication, it is always a simplified form.[8] Witness the iPhone and its like moving us further into the disembodied, truncated symbolic, as texting replaces voice-to-voice communication.

In 1945, Claude Lévi-Strauss commented that sociology, and by implication social anthropology as well, cannot explain the genesis of symbolic thought.[9] He may not have considered an effort made 30 years before by Freud's disciple and translator, Ernest Jones. One need not swallow all of Freud's metaphysics to draw upon his insights, beginning with Freud's supposition that symbolization is an unconscious process. He also held that repression establishes the unconscious, accessible only through dreaming. In his 1916 essay "The Theory of Symbolism," Jones proposed that "only what is repressed is symbolised; only what is repressed needs to be symbolised."[10] With the symbol, we moved away from immediacy and direct life. Jones called civilization "the

---

8   Sandra Wallman, "Appropriate Anthropology and the Risky Inspiration of 'Capability' Brown," in Alison James, et al., eds., *After Writing Culture* (New York: Routledge, 1977), p. 244.

9   Alan Barnard, *Genesis of Symbolic Thought* (New York: Cambridge University Press, 2012), p. 5.

10   Agnes Petocz, *Freud, Psychoanalysis, and Symbolism* (New York: Cambridge University Press, 1999), p. 14.

result of an endless process of symbolic substitutions."[11] His approach echoes Freud's premise that the repressed is sublimated into symbolic products as we trade away Eros and freedom for civilization.

The unconscious is structured like language, according to Lacan's well-known-formulation, and language regulates the unconscious, not the reverse.[12] So language becomes the subject, its constitutive role paramount. "Language speaks," said Heidegger.[13]

In *The Mechanization of Mind*, Jean-Pierre Dupuy concludes that thinking amounts to computations of representations.[14] Symbolic technologies have achieved a determining influence on human cognition. Personality disorders have accompanied this development: autism, schizophrenia, bipolar illness, obsessive-compulsive disorder, among others.

The Lord of the Rings by J.R.R. Tolkien is a tale of a mysterious power which, if possessed, promises empowerment—but ends instead in destruction and spiritual debilitation. An allegory of technology, very plausibly; perhaps more deeply, about the reign of representation.

"One Ring to Rule them, One Ring to Find them

One Ring to bring them all and in the darkness bind them..."[15]

---

11 R.H. Hook, "A Psychoanalytic Point of View," in R.H. Hook, ed., *Fantasy and Symbol: Studies in Anthropological Interpretation* (New York: Academic Press, 1979), p. 277.

12 Ellie Ragland-Sullivan, *Jacques Lacan and the Philosophy of Psychoanalysis* (Urbana, IL: University of Illinois Press, 1986), p. 101.

13 Martin Heidegger, *Heidegger: Poetry, Language, Thought*, translated by Albert Hofstadter (New York: Perennial Library, 1971), p. xxv.

14 Jean-Pierre Dupuy, *The Mechanization of Mind* (Princeton, NJ: Princeton University Press, 2000), p. 13.

15 See Christopher Tolkien, *The History of The Lord of the Rings* (Boston: Houghton Mifflin, 1988), p. 258, and Jane Chance, *The Lord of the Rings: The Mythology of Power* (Lexington, KY: University of Kentucky Press, 2001).

The book is complex and even convoluted at times, but I can't help seeing its main theme as related to symbolic thought.

Our encirclement by representation has trapped us in symbols, and keeps presence at bay. Wittgenstein said that no problem in philosophy can be solved until every philosophical problem has been solved.[16] Could it be that the solution to the puzzle of representation would show "every philosophical problem" in a radical new light?

---

16 Michael Dummett, *Frege: Philosophy of Language* (Cambridge, MA: Harvard University Press, 1981), p. 266.

ART AND MEANING

According to Joan Miró in 1927, "Art has been decadent since cave art."[1] He felt that painting hadn't developed much since its beginnings on the cave walls of the Upper Paleolithic. A deeper question has to do with art's very nature. It exists, arguably, out of a lack, as a substitute for what is missing. Yrjö Hirn referred to a general dissatisfaction and its counterpoint in art: "...the same longing for fuller and deeper expression which compels the artists to seek in aesthetic production compensation for the deficiencies of life."[2] It shouldn't be surprising that art has needed its defenders over the years.[3]

"The more fearful the world becomes, the more art becomes abstract," was Paul Klee's prescient remark.[4] This judgment reaches its fullness with Abstract Expressionism in the late 1940s and early 1950s, the high point of what is called modernism.

Critic Harold Rosenberg put this emphasis on it: "The new movement is, with the majority of painters, essentially a religious movement."[5] The Abstract Expressionists imagined themselves as outside capitalist culture, which they disdained, in solitary battle. There was, as Rosenberg saw, a redemptive aspect, a messianic-prophetic sense of responsibility to their efforts. This "heroic" aspect was soon to be

---

1   Catherine Grenier, *Big Bang: destruction et création dans l'art du 20me siècle* (bi-lingual) (Paris: Centre Pompidou, 2005), p. 84.

2   Yrjö Hirn, *The Origins of Art* (London: MacMillan and Co., 1900), p. 113. This is the main point of my "The Case Against Art" in John Zerzan, *Elements of Refusal* (Seattle: Left Bank Books, 1988).

3   For example, Christine Herter, *Defense of Art* (New York: W.W. Norton, 1938) and The Necessity of Art (New York: Penguin Books, 1959).

4   Darrell D. Davisson, *Art After the Bomb: Iconographies of Trauma in Late Modern Art* (Bloomington, IN: AuthorHouse, 2009), p. 113.

5   Harold Rosenberg, *Art and Culture* (Boston: Beacon, 1961), p. 41.

contested by those who were more interested in money, who were in no sense oppositional.

Mark Rothko produced deeply spiritual works consonant with his anarchist beliefs and aims. Mako Fujimura put it succinctly: "He painted the abyss."[6] Robert Motherwell's equally abstract *Elegies to the Spanish Republic* radiate a brooding, powerful intensity. With Jackson Pollock, a line is no longer a boundary, but a search for its own beginning or end. His large action paintings display an unparalleled primal, utopian energy. In an age of entropy, the Abstract Expressionists, each in his or her own way, strove to find a way out, to point beyond art via art.[7]

An important source or inspiration for many, including Rothko, Pollock, Clifford Still, and Adolph Gottlieb, was the art of several Native traditions—as antidote to modern decay and as renewal of creative capacities.

An opening to the non-built world also beckoned. In a well-known confrontation with older, established artist Hans Hoffman, Pollock declared, "I am nature." This was in response to Hoffman's verdict that Pollock's painting would run dry because he didn't work from nature.[8] His comeback meant that Pollock was part of nature, an enactment of nature. Elizabeth Langhorne writes perceptively of Pollock's dream of a Biocentric art, which would have involved, among other aspects, showing works unframed, out-of-doors.[9] This was unrealized, however; conceivably because such an effort

---

6   Eric Slade, *Rothko: Life Beyond the Abstract* (Portland, OR: video by Oregon Public Broadcasting, 2018).

7   See my "Abstract Expressionism: Painting as Vision and Critique," in John Zerzan, *Running on Emptiness* (Los Angeles: Feral House, 2002).

8   *Ibid.*, p. 102.

9   Elizabeth L. Langhorne, "Pollock's Dream of a Biocentric Art: The Challenge of His and Peter Blake's Ideal Museum," in Oliver A.I. Botar and Isabel Wunsche, eds., *Biocentrism and Modernism* (Burlington, VT: Ashgate, 2011).

would have served to underline, rather than to bridge, the gulf between culture and the natural world.

Any promise of radical change seems to have evaporated by the mid-1950s, as Pop Art arrived, hard on the heels of Abstract Expressionism. The era of the autonomous individual was over. Here was the first flush of postmodernism, when subjectivity and individuality go missing.

Although production for the market has been a fundamental condition of art since the Renaissance, Pop Art impudently embraced the language of commerce and the commodity. It became a part of the media system.

Gay artists such as Jasper Johns and his partner Robert Rauschenberg challenged not only the masculinist bias of AE, but also its notion of a lost deep meaning in need of recovery. Disillusion had set in, and resistance faded. Pop figures wanted to cooperate with the culture and its objects, rather than critique them. They were with the times, not against them: contemporary, detached, deeply complicitous. Any critical gestures were annulled by ambiguity and above all, by irony. Originality was explicitly rejected "in favor of a practice oriented to mediation and repetition."[10] Banal and superficial, e.g., the productions of Roy Lichtenstein, Jeff Koons.

Andy Warhol was a comparative latecomer to Pop Art's carnival of massified mediocrity, but he became its biggest star. "He paints the gamy glamour of mass society with the lobotomized glee that characterizes the cooled-off generation," wrote Robert Rosenblum.[11] From celebrities to soup cans, Warhol had begun in commercial advertising and never really left it. Fond of saying he wanted to be a

---

10 Stephen Melville, "Postmodernism and art," in Steven Connor, ed., *The Cambridge Companion to Postmodernism* (New York: Cambridge University Press, 2004), p. 89.

11   Robert Rosenblum, "Saint Andrew," *Newsweek*, December 7, 1964, p. 100.

machine, he aptly dubbed his studio The Factory. Pale and blank-faced, Warhol could not help but embody vapid and empty Pop output.

Many decades on, painting seems to have succumbed to the "moronic inferno of the information age," as Martin Gayford put it in a review by Matthew Brown, which pointed out that art "hasn't offered anything new for a long time now."[12] Postmodernism came in with Pop Art, and out went any notion of possible radical transformation of society.[13] Doubt-filled gestures, tentative and diffident, seem to typify today's art scene. In our thoroughly image-saturated and digitally mediated culture, the most pervasive idea is that no one *should* have a grasp of what contemporary art is.[14] This credo exemplifies the first principle of postmodernism: refusal of overview or metanarrative. And the fact of technological society is at the heart of this cultural state of affairs.[15] Artifice is pervasive, and art is lost in the shuffle—everywhere and nowhere. It lacks both development and a grasp of how much a part of the dominant order it has always been.

In his *Negative Dialectics*, Theodor Adorno decided that "The freedom of philosophy is nothing but the capacity to lend a voice to its unfreedom."[16] The same can be said of art.

---

12 Matthew Brown, "But Is It Tart?," *Times Literary Supplement*, October 19, 2018, p. 20.

13 A refreshing counter-perspective is Liam Dee's cheeky but informed *Against Art and Culture* (New York: Palgrave Macmillan, 2018).

14 Terry Smith, *What is Contemporary Art?* (Chicago: The University of Chicago Press, 2009), p. 1.

15 Lorenzo Simpson develops the idea that postmodernism is, at base, a function or outcome of technology in his very important *Time, Technology, and the Conversations of Modernity* (New York: Routledge, 1995).

16 Theodor Adorno, *Negative Dialectics* (New York: Continuum, 1997), p. 18.

NIGHT

Night is so often a stand-in for what is concealed, what is fearsome or evil. Night is even the realm of demons, incubi, black magic, nightmares, unseen dangers of all kinds—the negative side of reality. In this vein Joyce Carol Oates has the coming of night "the drawing of darkness out of the basic dark of the world."[1]

On a personal and contemporary level, that "3 a.m. feeling" is well-known, in a world where so many fear being alone. In an Age of Distraction, night—unadorned—offers fewer diversions. For better or worse, night is more conducive to being in the here and now. Henry Beston noted in the 1920s that our "civilization has fallen out of touch with many aspects of nature, and none more completely than with night."[2]

"But how do you understand the night,"[3] asks Heidegger, adding, "In the dark I see nothing, and nonetheless I see."[4] Gaston Bachelard felt that nights are singular, unrelated, without history or future.[5] Much remains to be explored, however. In fact, in the same work, *The Poetics of Reverie*, Bachelard wants to know, "Where is the philosopher who will give us the metaphysics of the night, the metaphysics of the human night?"[6]

---

1 Joyce Carol Oates, *Anonymous Sins & Other Poems* (Baton Rouge: Louisiana State University Press, 1969), p. 37.

2 Henry Beston, *The Outermost House* (Garden City, NY: Doubleday, 1928), p. 165.

3 Quoted in David Michael Levin, *The Opening of Vision: Nihilism and the Postmodern Situation* (New York: Routledge, 1988), p. 374.

4 *Ibid.*, p. 373.

5 Gaston Bachelard, *The Poetics of Reverie* (Boston: Beacon Press, 1971), p. 145.

6 *Ibid.*, p. 147.

There is sublimity—also, at times, terror. "Take Back the Night" is an important reminder of the latter. Night can be an unsettling staring into the void, or an invitation to immersion and participation. The poet Novalis pondered these aspects, wondering, "Are you teasing us, dark Night? What're you holding under your cloak, that grabs so unseen at my soul?"[7] His "Hymns to the Night" proclaim night's gifts and prescribe openness to the coming of a deep, non-rational experience. All real knowledge arrives as a gift.

The darkness at night is Earth's shadow, Earth blocking the sun. Composer Franz Schubert referred to it quite differently as he set to music Johann Gabriel Seidl's poem "Bright Night." He expressed night in this piece as bathed in "rich light." Stevie Smith's "The Light of Life" enjoins us to "Put out that Light,/Put out that bright Light,/ Let Darkness fall."[8] In the night's vast stillness is a darkness that can be felt. In the nocturnal deep (battling light pollution and noise pollution, of course) is an undeniable presence.

The 18th-century poet Edward Young proved himself night's worthy laureate. His *Night Thoughts* found the dimension freeing and profound, liberated from the bonds of daytime restraints. Lorus and Margery Milne's *The World of Night*, two centuries later, sees the dark world as a place new and undiscovered. Speaking of exploration and discovery, David G. Campbell recorded an Arara woman in the '90s in the Amazon: "The comet stretched across half the sky, night after night I came to the edge of the river and watched. Now that was discovery!"[9]

---

7   *Novalis: Hymns to the Night*, translated by Dick Higgins (New Paltz, NY: McPherson & Company, 1984), p. 11.

8   *Stevie Smith: Collected Poems* (New York: New Directions, 1983), p. 372.

9   David G. Campbell, "The Explorer's Journey," in William H. Shore, ed., *The Nature of Nature* (New York: Harcourt Brace & Company, 1994), p. 29.

When all of day's roads seem to lead nowhere, the night may offer aid. Let the flower of night unfold for us. Byron:

> And this is the Night — Most glorious Night!
> Thou wert not sent for slumber! let me be
> A sharer in thy fierce and far delight, —
> A portion of the tempest and of thee![10]

There is potential in David Michael Levin's counsel "that we be open to the radical and essentially subversive teachings of the night."[11] Night's hush and invitation to solitude attract those who seek a haven from stress and from conformity.

Henry Vaughn's "The Night" opens with "Dear Night! this world's defeat,/ The stop to busy fools, care's check and curb."[12] It has been a time of rest and of reflection.

And many people, in Europe for example, became more powerful at night, in de facto control of the nocturnal landscape. "Midnight feastings are great wasters,/ Servants' riots undo masters," according to an Elizabethan saying.[13] Despite the steadily rising powers of the early modern state, "nighttime defied the imposition of government authority."[14] Gatherings of all sorts took place after work and sunset, the camaraderie often stoked by ale, beer, or wine. Spinning or knitting bees could be the occasions for female solidarity, away from men.

---

10   "Child Harolde's Pilgrimage," *Byron's Poetry*, Frank D. McConnelly, ed. (New York: W.W. Norton, 1978), p. 71.

11   Levin, *op. cit.*, p. 349.

12   Henry Vaughn, "The Night," in *A Treasury of Great Poems*, Louis Untermeyer, ed. (New York: Simon & Schuster, 1942), p. 491.

13   Quoted in A. Roger Ekirch, *At Day's Close: Night in Times Past* (New York: W.W. Norton, 2005), p. 255.

14   *Ibid.*, p. 88.

From the late Middle Ages, an assault on the autonomy of the submerged classes took the form of a profound demonization of night. It was assailed as the realm of witches and other agents of Satan. With the 18th century the effort to reclaim the night took a new form: public illumination. Lighting had great promise as a weapon of social control—and for this reason "where urban disorders occasionally flared, among the first casualties were street lamps."[15] People were also hung from lampposts—the French rallying cry was "à la lanterne!"

Gas lighting was developed in the context of the Industrial Revolution; it enabled round-the-clock factory operations. Night became subject to industrialization and the subsequent consumerism. There is a very sizable literature on the increasing toll, socially and personally, as nighttime is ever more invaded and colonized.[16] Less sleep and a rise in sleep disorders are among widespread and worsening problems. Steadily diminished night also has profound impact on ecological systems and patterns. No less than our oldest path to the human psyche is being obliterated.

Various religious traditions have valorized the night. The Koran is based on Mohammed's nightly prayers, his "nightly journeys." Sixteenth-century Spanish mystic St. John of the Cross, in his *Dark Night of the Soul* and elsewhere, describes night as "a friend, even the supreme friend," in George Tavord's words.[17] *Dark Night of the Soul* contains this typical exclamation: "O guiding night!/ O night more lovely

---

15 *Ibid.*, p. 336.

16 For example, Martha Gies, *Up All Night* (Corvallis, OR: Oregon State University Press, 2004); Murray Melbin, *Night as Frontier: Colonizing the World After Dark* (New York: The Free Press, 1987); Martin Moore-Ede, *The Twenty-Four-Hour Society* (Reading, MA: Addison Wesley, 1993).

17 George H. Tavard, *Poetry and Contemplation in St. John of the Cross* (Athens, OH: Ohio University Press, 1988), p. 58.

than dawn!"[18] Eastern spirituality in general, and Taoism in particular, are even more consistently and more deeply at home in this dimension. "Darkness within darkness,/ The gateway to all understanding," proclaims the Tao Te Ching.[19]

In terms of sensory understanding, night reminds us of an acuity of the senses that we once had, and may have again. Hearing, touch, and smell necessarily re-assert themselves as seeing recedes in importance. "The day has eyes, but night has ears," according to a Scottish proverb. Hermia in *A Midsummer Night's Dream* contemplates "Dark night, that from the eye his function takes,/ The ear more quick of apprehension makes./ Whereupon it doth impair the seeing sense,/ It pays the hearing double recompense." But it is also true that even without a moon, a starlit sky enables humans and many other animals to see quite well. Once the pupils have widened and the retinas have adjusted, a person can see almost as well as an owl or a lynx.[20] Peripheral vision may actually sharpen.[21]

Dusk announces the time for philosophy when, as Hegel saw it, the Owl of Minerva, the goddess of wisdom, takes flight. Sunset is not tame; night does not fall, but rises. The dominant daytime world begins to appear less believable. Novalis observed, "How poor and childish the Light seems now—how happy and blessed the day's departure."[22] D.H. Lawrence's "Twilight" shares a similar sentiment: "All that the worldly day has meant/ Wastes like a lie."[23] And so to

---

18   *Ibid.*, p. 57.

19   *Tao Te Ching*, new English version by Stephen Mitchell (New York: Harper & Row, 1988), p. 1.

20   Melbin, *op. cit.*, p. 8.

21   Ekich, *op. cit.*, p. 124.

22   Novalis, *op. cit.*, p. 13.

23   "Twilight," in *The Complete Poems of D.H. Lawrence*, Vivian de Sola Pinto and Warren Roberts, eds. (New York: The Viking Press, 1964), p. 41.

cast off into the stillness of the night; the darkness thickens, opens to essentials.

In the city searchlights poke up through the night, as if searching for something. In the wordless star-lashed dark, the night crying its truth, as ever. One sun by day, ten thousand shine by night, Steady Pole Star, glittering Pleiades, Orion's studded belt... Overflowing heavens of stars in a nightscape old as water, older. The dotted waves of night, always breaking. "Tonight the stars are like a crowd of faces/ Moving round the sky and singing/ and laughing," wrote Wallace Stevens.[24] Byron found "the language of another world" in "night's starry shade/ Of dim and solitary loveliness."[25] John Hollander provides some lines of promise: "The world is everything that happens to/ Be true. The stars at night seem to suggest/ The shapes of what might be."[26]

"Come up, thou red thing./ Come up, and be called a moon" was D.H. Lawrence's address in his "Southern Cross."[27] Emblems of wholeness, face of the moonful midnight sky. Lunar mythology seems universally to have preceded solar mythology.

"Hear the night bellow,/ our great black bull. Hear the dawn/ distantly lowing," Hayden Carruth perceives.[28] Night in retreat, not wanting to "face the music"... dawn. Hart Crane expresses reluctance to surrender: "Serenely now, before day

---

24   "Dezembrum," in *The Collected Poems of Wallace Stevens* (New York: Alfred A. Knopf, 1971), p. 218.

25   "Manfred," in McConnell, ed., *op. cit.*, p. 155.

26   John Hollander, "The Great Bear," in *The Norton Anthology of Modern Poetry*, Richard Ellmann and Robert O'Clair, eds. (New York: W.W. Norton, 1973), p. 1241.

27   D.H. Lawrence, "Southern Night," in de Sola Pinto and Roberts, eds., *op. cit.*, p. 302.

28   "Dawn," in *The Selected Poetry of Hayden Carruth* (New York: Macmillan, 1985), p. 45.

claims our eyes/ Your cool arms murmurously about me lay."[29] Night consoles, but must give way to day. Novalis again: "Now you, bright light, are waking those tired ones to work."[30]

In the night's less structured qualities insight may lurk, a flash of lightning, in more ways than one. As Rilke put it, "it is possible a great energy/ is moving near me./ I have faith in nights."[31] The thing can go either way. Iago (*Othello*, Act 5, Scene 1) puts it bluntly: "This is the night; that either makes me or does away with me quite." Night is really the last place to hide.

Emmanuel Levinas saw in night "the very experience of the *there is*... a presence, an absolutely unavoidable presence."[32] Once we gathered in the night around the fire, unmediated, surrounded by the universe, listening and learning. Green and present is the night.

> "She walks in beauty, like the night..."
> —Byron

---

29 "The Harbor Dawn," in *The Complete Poems of Hart Crane*, Marc Simon, ed. (New York: Liveright, 2000), p. 53.

30 Novalis, *op. cit.*, p. 21.

31 Untitled, *Selected Poems of Rainer Maria Rilke*, translated by Robert Bly (New York: Harper & Row, 1981), p. 5.

32 Emmanuel Levinas, "From existence to ethics," in *The Levinas Reader*, Sean Hand, ed. (New York: Blackwell, 1989), p. 30.

DEATH

Sooner or later each of us ceases to be. A living organism is one that will die. All life comes to an end. And not just everything living but everything. The Second Law of Thermodynamics decrees doom for everything material. Entropy implies the death of the universe.

How brief and taunting is life, how deep the sea. Death is seldom timely; it often comes too soon and sometimes too late. Does this impermanence, this finitude render life meaningless? For some, life lacks significance if it cannot be objectified and made permanent somehow to memorialize its mortal inhabitant. After all, it is often asserted, the fear of death is the motivation of all human endeavors.

Our mortality may not rob life of meaning, but may instead prompt us to find meaning in life. To go further, could there be meaning without death? Doesn't life have meaning precisely because of death? It intensifies life. The poet Wallace Stevens finds in death "the mother of beauty."[1]

Socrates famously proclaimed, "The one aim of those who practice philosophy in the proper manner is to practice dying and death."[2] In his *Tusculan Disputations*, Cicero likewise said philosophizing is nothing but consideration of death, making ready for death. No one approaches philosophy that way today, but a fundamental challenge remains—the most fundamental of all.

At the same time we know how difficult it is to cope with the thought of our own death. As Rousseau put it, "He who pretends to look on death without fear lies."[3] We go to great

---

1   Wallace Stevens, *Collected Poems*, "Sunday Morning" (New York: Knopf, 1971), p. 69.

2   Plato, *Phaedo* (New York: Cambridge University Press, 1993), 64a.

3   In Jean-Jacques Rousseau's *Julie, or the New Heloise*, quoted in D.J. Enright, ed., *The Oxford Book of Death* (New York: Oxford University Press, 1983), p. 22.

lengths to hide from death. We refer to the "dear departed"; people "pass on," or simply "pass." An increasingly tech- nologized medical terminology has patients "terminating" rather than dying, but blunter words are in constant use, usually with a different, safer referent. For example, a base-runner may "die on base," or a football game culminate in a "sudden death" overtime.

Thinking about death seems to defy thought itself. It is elusive in the sense that only negations come to mind, only what is not there. And of course, the real elephant in the room is our own death. Aphorist La Rochefoucauld noted, "One cannot look directly at the sun or at death."[4] As civiliza- tion makes our lives ever more circumscribed, death becomes scarier. The swindle is felt, if not articulated. Is this all we get? Having control of so little in our lives, we don't even see death as something of our own. "The desire for a death of one's own is growing more and more rare. In a little while it will be as rare as a life of one's own," Rilke wrote.[5] Octavio Paz added, "Nobody thinks about death, about his own death, as Rilke asked us to, because nobody lives a personal life."[6]

Anticipation of our death may reveal all too much about what we've been allowed to be. We're not born alone, but we may die alone. More and more of us live alone, and more die, alone, in the technologized hospital that rules death. Grief lurks behind the death culture of technological illusions and separation. Loss pervades our lives.

My guess is that's what drives the transhumanists' wor- ship of technology as salvation, in particular their claim that death can be abolished. Immortality is the Holy Grail

---

4   Quoted in Zygmunt Bauman, *Mortality, Immortality, and Other Life Strategies* (Cambridge, UK: Polity Press, 1992), p. 15.

5   Rainer Maria Rilke, "The Notebooks of Malte Laurids Brigge," in *Rainer Maria Rilke: Prose and Poems* (New York: Continuum, 1984), p. 6.

6   Octavio Paz, *The Labyrinth of Silence* (New York: Grove Press, 1985), p. 57.

promised by transhumanists Ray Kurzweil and Zoltan Istvan. Madly delusional, their doctrine shows that the more "advances" in technology, the more fully technology dominates society, the more fear and denial of death is evident.

Turning to the other end of the spectrum, we encounter animals and early human animals. It has been asserted as a given, from Socrates to Freud et al., that a cardinal difference (perhaps *the* cardinal difference) between non-human animals and ourselves is that the former are unaware of their mortality. But how do we really know this? Françoise Dastur questions the assumption: "it is less than evident that animals have no presentiment whatsoever of their death."[7] And as noted above, are we humans even capable of grasping the datum of our death?

For early humans, the general idea of death is another matter. Given that *Homo* had intelligence equal to ours a million years ago, death must have been plainly observable to them; for Plinio Prioreschi to aver that "the abstract concept that death could occur to all... must have happened quite late, probably at the very beginning of the dawn of civilization,"[8] is absurd, a baseless conceit of the civilized. Similarly, Zygmunt Bauman's idea that only with the appearance of graves did we cross "the threshold of humanhood"[9] is also misplaced. Early humans almost certainly buried their dead. Sanitation alone required some form of disposal/ burial; there is no necessity to impute symbolic thinking to the practice. "Burial rites" is an oddly popular term in the literature about prehistory, conflating two distinct things. Burials could and likely did take place without rites or

---

7   Françoise Dastur, *Death: An Essay on Finitude* (Atlantic Highlands, NJ: Athlone, 1996), p. 6.

8   Plinio Prioreschi, *A History of Human Responses to Death* (Lewiston, NY: The Edwin Mellen Press, 1990), pp. 19–20.

9   Bauman, *op. cit.*, p. 51.

ritual—that is, without projecting the symbolic into the practice. A symbolic approach to the cycle of birth and death only definitively entered the picture with agriculture, in the last 10,000 years.

Death arrives as a private and individual event, intimate and incommunicable. It is not abstract; it is seen very differently in different times, places, and cultures. The Aztecs seem to have worshipped death; Egyptians associated death with immortality, whereas their sometime enemies the Hittites and Mesopotamians rejected the afterlife completely. Greek culture in the eighth through sixth centuries B.C. strongly exhibits an acute death-consciousness, while a bit later, the Athenian Epicurus argued that death is an irrelevancy, nothing to be worried about. The Jewish tradition transfers immortality from the individual to community memory, whereas Christianity and Islam stress personal immortality after death.

Through most of the Middle Ages death was a familiar and relatively public event in an individual's life. Many were present at the deathbed; as late as the 17th century, portrayals of deathbed scenes included children.[10] By the 18th century, while certain themes and rituals persisted, death was being furtively pushed out of the world of familiar things and was becoming progressively shameful and hidden.

Enlightenment denial of the immortality of the soul became the standard philosophical position on death. For Hegel, overcoming death was pivotal to a full life. In the best-known pages of his *Phenomenology of Spirit*, dealing with the master-slave dialectic, Hegel argues that only by confronting "Lord Death" can we fully develop. Death drives thought, and plays a key role in the formation of Spirit—key to the entire Hegelian system.

---

10 Philippe Ariès, *Western Attitudes toward Death: From the Middle Ages to the Present* (Baltimore, MD: The Johns Hopkins University Press, 1974), p. 12.

His friend, lyric poet Friedrich Hölderlin, focused heavily on death; it pervades the pages of his texts. One recurring motif is characters taking their own lives out of a sheer plenitude of existence. "Once I lived like the gods, and more is not needed," he wrote.[11] The protagonist of *The Death of Empedocles* declares, "For death is what I seek. It is my right."[12]

The bitter pessimism of Arthur Schopenhauer (*The World as Will and Representation*, 1818) pronounced suffering and death the central aspects of existence, the latter being the real aim of life. But by mid-century, philosophy largely avoided the subject of death. Hegel and Hölderlin were dead before 1850, and in society the deathbed retreated further from view.

Friedrich Nietzsche came to reject the death fixation of Schopenhauer and of Richard Wagner as decadent and unhealthy. He had been enraptured with both of them early on, especially with Wagner. Wagner's music celebrated what he saw in Schopenhauer, "the genuine ardent longing for death."[13] This was exactly what Nietzsche came to oppose.

The seductiveness of dissolution is pre-eminent in Wagner's operas, and of course can be found in such early-20th-century works as Joseph Conrad's *Heart of Darkness* and Thomas Mann's *Death in Venice*.

Sigmund Freud elevated the appeal of the end into a psychological axiom, the death instinct. In physics, the goal of any energy process is a state of rest. There are many versions of an analogous, supposedly universal human yearning for primordial unity, a return to a simpler state. But

---

11   Friedrich Hölderlin, "To the Fates," quoted in Walther A. Kaufman, *The Faith of a Heretic* (Princeton, NJ: Princeton University Press, 2015), p. 370.

12   Hölderlin, *The Death of Empedocles*, quoted in Leslie Hill, *Blanchot: Extreme Contemporary* (New York: Routledge, 1997), p. 245.

13   Richard Wagner, *Wagner on Music and Drama* (New York: Da Capo Press, 1988), p. 270.

Freud took this much further, claiming that a death drive is central to all life, down to the cellular level. This has seemed hard to follow, even for many Freudians. It seems arbitrary; its late appearance may be due to dark circumstances in Freud's life—the death of his daughter and a grandson, his worsening cancer of the jaw, not to mention the enormity of World War I carnage and rising anti-Semitism in Europe.

Martin Heidegger brings a renewed philosophical emphasis on death through his defining attention to what it means to *be*. Death as the determining factor of selfhood is a key aspect of his *Being and Time* (1927). Our existence is complete only in view of its end; we are free only when embracing our finitude. Not only Heidegger, but existentialism in general had a striking preoccupation with death and its challenge.

As the horrors of contemporary civilization multiply, in an increasingly atomized society, writers largely respond to the idea of death with sadness and despair. Throughout industrialized society there is forgetfulness of death. Where more community and familial activity survives (in Italy, Spain, and Mexico, for example), people don't flee from death so readily. Philippe Ariès concluded his major historical study of death with the finding that in "the most industrialized, urbanized, and technologically advanced areas of the Western world... society has banished death."[14]

Herbert Marcuse was one of the rare thinkers who explored the sociopolitical dimensions of death, agreeing with Karl Jaspers that "death as an objective fact of existence is not itself a limiting situation."[15] Marcuse condemned death-obsessed Western philosophy as missing the relevance of fundamental realities of social existence.

14 Philippe Ariès, *The Hour of Our Death* (New York: Knopf, 1981), p. 560.

15 Herbert Marcuse, "Human Death," quoted in José Ferrater Mora, *Three Spanish Philosophers: Unamuno, Ortega, Ferrater Mora* (Albany, NY: State University Press of New York, 2003), p. 176.

As death became more depersonalized in mass society, the sense that we die alone became ever more deeply instilled in us. Thus the natural fact of death becomes denial of death, and partakes of denial more broadly. No domination is complete without the threat of death, and in this way death causes an extra degree of anxiety, within the realm of unfreedom that is domestication/civilization.[16] Shakespeare's *Measure for Measure* dramatizes death at work as an ideology of social control.

Marcuse asserts that in a better world, death could be reduced to its biological reality; the tragic separation of death from life would be greatly altered. Anxiety and denial might be much reduced in a healthy, free context. The natural fact of death need not be a social institution.

Writing about death while feeling healthy is a safe endeavor, one I can approach with equanimity or evenness of mind. It might come out somewhat differently, in a less detached manner, if it were now otherwise for me.

I've had my lows and highs, but mainly have been able to pursue what I've wanted to pursue. It's obvious that not everyone has had that privilege.

Death is the ground and condition of our lives. There is a testing that comes to an end, a kind of responsibility owing to our finiteness. Each of us is irreplaceable, which magnifies the responsibility. "Become what you are," challenged Nietzsche, echoing Pindar.[17] Some counselors ask their clients to write their own obituaries, a summing-up as if there is no longer anything that could be changed. What

---

16 Marcuse argued in *Eros and Civilization* that we could have civilization without its excess of repression. Given that civilization exists because of domestication—control at a fundamental level—his argument fails. Somewhere he asked how past suffering might be redeemed. Decades after reading this question, its audacity still moves me. Ending civilization would put suffering in a new perspective, in a redemptive light.

17 Dastur, *op. cit.*, p. 66.

was it all about? A person with unlived lives is much more apt to fear death. One who has never properly lived has no proper time to die.

"Death opens the door to metaphysics. We must be bold and step through it," according to Palle Yourgrau.[18] On a slightly less high-flown plane, Walter Benjamin observed that "Death is the sanction of everything the storyteller can tell. He has borrowed his authority from death."[19] Dying simply keeps us from having more of a good life; so thinking about death leads to thinking about what a good life means to us. Quite possibly that boils down to the sheer pleasure of going on living, to see what happens next!

We may be aware that time is just another take on the force field of estrangement, just another word for it. But in the usual sense of the word, the clock keeps on ticking, often entering without knocking. In André Malraux's *The Royal Way*, Claude says in terror, "There is... no death.... There's only... I... I... who am dying."[20] The promise of the challenge endures: "But for your Terror/ Where would be Valour?" asked Oliver St. John Gogarty.[21]

The famous death agony in Tolstoy's *The Death of Ivan Ilyich* is about terror, and how it is undone by love. His pain at death's door melts away when he feels the love of a son and a servant. As Gabriel Marcel put it, "to love a person is to say: 'you shall not die.'"[22] How many times do we hear of couples never apart for many years, who when separated die within days of each other?

---

18 Palle Yourgrau, "The Dead," *Journal of Philosophy* 86 (1987), p. 84.

19 Walter Benjamin, *The Storyteller: Tales out of Loneliness* (New York: Verso, 2016), p. 7.

20 André Malraux, *The Royal Way* (New York: Smith and Haas, 1995), p. 290.

21 Oliver St. John Gogarty, "To Death," quoted in Enright, *op. cit.*, p. 40.

22 Quoted in Prioreschi, *op. cit.*, p. 30.

Dylan Thomas praised his father's final fury in the poem that ends "Do not go gentle into that good night,/ Rage, rage against the dying of the light."[23] And yet, as many doctors have reported, most of their patients know they are going to die, and most of them are ready.[24]

Woody Allen: "I don't mind the idea of dying. I just don't want to be there when it happens."[25] But seriously, folks. I think Samuel Johnson had it right when he decided, "It matters not how a man dies, but how he lives. The act of dying is not of importance, it lasts so short a time."[26] Another way to say it is that there can be "a fate worse than death."

Will there be a final ease, a final understanding? These lines from D.H. Lawrence may be apropos:

> Swings the heart renewed with peace
> even of oblivion.
> Oh build your ship of death. Oh build it!
> for you will need it.
> For the voyage of oblivion awaits you."[27]

23 Dylan Thomas, "Do Not Go Gentle into That Good Night," in *The Poems of Dylan Thomas* (New York: New Directions, 2003), p. 162.

24 Arnold A. Hutchnecker, "Personality Factors in Dying Patients," in Herman Feifel, ed., *The Meaning of Death* (New York: McGraw-Hill, 1965), p. 238.

25 Quoted in Alfred G. Killea, *The Politics of Being Mortal* (Lexington, KY: University of Kentucky Press, 1988), p. 1.

26 Quoted in Enright, *op. cit.*, p. 53.

27 D.H. Lawrence, "The Ship of Death," in *The Complete Poems of D.H. Lawrence* (New York: Viking Press, 1964), p. 960.

MEANING IN THE AGE OF NIHILISM

In 2000 I wrote a few hundred words on "The Age of Nihilism." Some years later that descriptive title is far more apt than before. Like a poison gas settling down over a battlefield, nihilism has, at least for the moment, begun to cloak or threaten so much of contemporary life and thought. The old "meaning of life" question seems to have a new urgency, unless it's too late to explore it. Adorno provides a stunning response to the meaning of life/point of living query: "A life that has any point would not need to inquire about it. The question puts the point to flight."[1]

A pointless life is a relatively new phenomenon. Terry Eagleton poses the meaning-of-life question in the context of 19th-century British literature, observing that a shift occurred in about 1870. Before that date, writers like Jane Austen and William Thackeray rarely referred to it, whereas after 1870 Thomas Hardy, Joseph Conrad, and others addressed it with some urgency.[2] Unsurprisingly, this cultural turning point coincides with the decisive ascendancy of industrial life in England (a worldwide first). At base, today's meaninglessness is a function of machine existence. Viktor Frankl, referring to the individual as "a being in search of meaning," goes on to note that "today his search is unsatisfied and thus constitutes the pathology of our age."[3]

A contagious nihilism accompanies the crisis all around us. The absence of meaning and value is seen in rising suicide rates, epidemic addictions, suicide bombings, and rampage shootings, and so much else in the landscape of no

---

1   Theodor W. Adorno, *Negative Dialectics* (New York: Continuum, 1973), p. 377.

2   Terry Eagleton, *The Meaning of Life* (New York: Oxford University Press, 2007), p. 34.

3   Quoted in Dennis Ford, *The Search for Meaning* (Berkeley: University of California Press, 2007), p. 12.

community. Because social life is based on the meaning it provides for its participants, social life itself is visibly waning.

Albert Camus famously opens *The Myth of Sisyphus* with the question of suicide: whether life is worth living. Many have the means to live, but no meaning to live for. The columnist David Brooks cites this example: "We ask students to work harder and harder while providing them with less and less of an idea of how they might find a purpose in all that work."[4]

Freedom is a struggle for meaning which is itself a dynamic force, according to William James, Viktor Frankl, and others. The drive to find or create meaning and value is a perpetual task, and at the heart of every human endeavor. In a mainly non-alienated world, this effort was possibly far less needed, compared with our civilization, where experience is de-grounded. We no longer feel at home in this massified, mediated world. We talk so much about meaning because meaninglessness is far advanced into all our lives. Yet it's very hard to confront. A common response is nervous laughter at Woody Allenish jokes or Monty Python's *The Meaning of Life*.

The meaning of life is meaning, but just where do we find that? All this varies hugely on the individual level. Women and men may see their own lives very differently from the inside. What we value can change at various points in our lives. Millions of self-actualization books and videos are sold; some check out Quality of Life approaches, or flock to hear the vague nostrums of the Dalai Lama. The approved idea is that the meaning of life is primarily an individual or personal affair—a severely limited, failed orientation.

Philosophy centers on the question of meaning; but this, too, has failed. As countless philosophical works attest, many if not most philosophers conflate the question of meaning

---

4   David Brooks, "Inside Student Radicalism," *New York Times*, May 27, 2016, p. A21.

with a close study of language and how it works.[5] During the past century, it has been generally acknowledged that meaning can't be considered apart from language; meaning is widely viewed as essentially a linguistic phenomenon. Meaning can only be approached and encompassed by means of one's (preferred) semantic theory; this method reduces philosophy to questions of signification, such as, how can we be sure what a given sentence says?

Philosophers such as Gottlob Frege, Michael Dummett, and Ludwig Wittgenstein dealt with meaning only in a narrow, formal sense, suggesting that we abandon it as misleading and study instead the way language is used. But Jan Zwicky, also a philosopher (and poet), avers that Wittgenstein never abandoned "the intuition that deep matters of value in some way fall outside the scope of language."[6] She goes on to say that "a significant part of the meaning of words rests in wordlessness."[7]

Language itself is value-laden; grammar calls the tune in fundamental ways, as I've discussed elsewhere.[8] Whatever can be represented can be controlled. Poet Stéphane Mallarmé responds to this constraint exquisitely: "Meaning is a second silence deep within silence; it is the negation of the world's status as a thing. This ever unspoken meaning which would disappear if one ever attempted to speak it..."[9] This unspoken,

---

5   e.g., Stephen R. Schiffer, *Meaning* (Oxford: Clarendon Press, 1972); Ruth Garrett Millikan, *Varieties of Meaning* (Cambridge, MA: The MIT Press, 2004); Stefano Predelli, *Meaning Without Truth* (New York: Oxford University Press, 2013); Vincent Descombes, *The Institutions of Meaning* (Cambridge, MA: Harvard University Press, 2014).

6   Jan Zwicky, *Lyric Philosophy* (Toronto: University of Toronto Press, 1992), thesis 118.

7   *Ibid.*, thesis 261.

8   "Language: Origin and Meaning," in John Zerzan, *Elements of Refusal* (Columbia, MO: C.A.L. Press, 1999); "Too Marvelous for Words," in Zerzan, *Twilight of the Machines* (Port Townsend, WA: Feral House, 2008).

9   Quoted in Peter Schwenger, "The Apocalyptic Book," Mark Dery, ed., *Flame Wars* (Durham, NC: Duke University Press, 1994), p. 64.

lived meaning defies reification and language games. Foucault was wrong: not all life is built around language.

Life takes place in this world, not on a page or a screen. It isn't fully possible to live the right life in a wrong world, even if we assert that the meaning of life is a life of meaning. In this destructively disenchanted context, some of us find meaning in commitment to a project, a goal of liberated life. Studies and surveys continually point to the obvious, that those who lack meaning tend to be less happy with every aspect of their lives.[10] Meaning has a physical aspect. There is a healing force there.

It is also valid to realize that meanings must be sought and found by each of us, in an active engagement with life. Relationship is the greatest single source of well-being, and the key to overcoming nihilism. Love and friendship are likely the most powerful motivators to do anything at all. Life cannot be without meaning to anyone who loves. Conversely, the nihilist does not change diapers.

It's said that the meaning of it all is to be found in moments. I think of how a stranger's smile can light up my day. To mean is to be present, present to the moments, transparent to presence. The recognition of what is important for its own sake. Meaning is irreducible. *All* of this is the meaning.

This is not the right world and everyone knows it. There is a longing, a need, and as Adorno has it, "the need in thinking is what makes us think." And this "need is what we think from, even where we disdain, wishful thinking."[11] Mark Rowlands puts it well: "Any satisfying account of the meaning of life must be capable of redeeming life."[12]

---

10 e.g., J.L. Freedman, *Happy People* (New York: Harcourt Brace Jovanovich, 1978).

11 Adorno, *op. cit.*, p. 408.

12 Mark Rowlands, *Running with the Pack* (New York: Pegasus Books, 2013), p. 100.

THE CASE AGAINST PHILOSOPHY

Philosophy is thinking at the most general level. It addresses questions or problems theoretically, abstractly. Philosophical knowledge is knowledge attained by avoidance of what is singular, non-generic. It is disembodied thought, decontextualized, removed from ordinary surroundings. As Agnes Heller put it, "philosophical activity is only possible via the suspension of particularity."[1]

Philosophy is an essentially impersonal pursuit or inquiry. Thus it is misleading to assert, as do Lakoff and Johnson, that "philosophic theories are attempts to make sense of our experience."[2] How can we plumb the human condition by using a method that is purposefully abstracted from the substance of life? William Desmond's caveat is ingenuous: "By its nature philosophy risks being merely abstract thought."[3]

Philosophy is allegedly above all a love of wisdom. What wisdom, whose wisdom? A search for the meaning of life? When was it lost?

Aristotle famously claimed that philosophy began in wonder, but his "wonder" avoided most of reality. A key reason for this avoidance lies in the quest for what is universal. Particularity, including individuals, is trampled underfoot by supposed universality. National differences in preferred philosophic approaches run counter to claims of philosophy's universality. The Anglo-American tendency, for example, is predominantly analytic/empiricist/pragmatic, and likely to refer to the more speculative efforts of French and German philosophers as unintelligible nonsense.

---

1   Agnes Heller, *Everyday Life* (Boston: Routledge & Kegan Paul, 1984), p. 111.

2   George Lakoff and Mark Johnson, *Philosophy in the Flesh* (New York: Basic Books, 1999), p. 337.

3   William Desmond, *Philosophy and Its Others* (Albany, NY: SUNY Press, 1990), p. 209.

There is certainly wide disagreement among philosophers. How many would claim progress or positive results overall, in the field? Blackford and Broderick tell us that the most debated "subject" in this century has been philosophy itself, accompanied by criticisms of devastating scope.[4] So much critical reflection, and has it helped?

Philosophy arises from negative experience, from encountering the world as "radically defective, disappointing, or unsatisfactory," in Raymond Guess' words[5]—a common and valid judgment. William James, on several occasions, noted that philosophy begins with discontent. There is a "need in thinking," as a book about Theodor Adorno phrased it.[6] We seem to have arrived at the fullness of this condition. After thousands of years of history, as Max Scheler wrote about a century ago, our age is the first to have been seen as wholly and completely problematic.[7]

And philosophy has utterly failed to make sense of the world, of reality. The opening line of Adorno's *Negative Dialectics* proclaims: "Philosophy, which once seemed obsolete, lives on because the moment to realize it was missed."[8] Hence the continuing, ruthless self-criticisms. There is little mystery to this situation. Philosophy arose from division of labor, and like every aspect of civilization, follows its alienating logic. Philosophy became a profession and an academic "discipline" in due course, along the road to hyper-specialization and self-absorption. Philosophy as

4   Russell Blackford and Damien Broderick, eds., *Philosophy's Future: The Problem of Philosophical Progress* (Hoboken, NJ: Wiley Blackwell, 2017), p. xv.

5   Raymond Guess, *Morality, Culture, and History: Essays on German Philosophy* (New York: Cambridge University Press, 1999), p. 79.

6   Donald Burke et al., eds., *Adorno and the Need in Thinking* (Toronto: University of Toronto Press, 2007).

7   Eugene Trias, *Philosophy and Its Shadow* (New York: Columbia University Press, 1983), p. 105.

8   Theodor Adorno, *Negative Dialectics* (New York: Continuum, 1997), p. 3.

delimited, and limiting. The literature is self-referential, situated determinately away from effort to grasp the actual, much less the whole.

The fundamental limitation is that of the symbolic itself. A "realized" philosophy would not be a practice of representation, but of presentation—a place of wholeness, immediacy, presence. Gilles Deleuze pointed out that philosophy remains caught in representation; at times he attacked this mode of thinking, in favor of "a theory of thought without images."[9] But it's hard to imagine such a move away from philosophy as a marketplace of symbolic goods. Symbolic culture may be the most durable facet of civilization, the most difficult for us to abandon.

Philosophy does not help with this challenge. In fact, philosophy has been going in precisely the wrong direction, down the rabbit hole of the symbolic. In the 20th century, language became its central theme or subject, from Gottlob Frege to J.L. Austin and A.J. Ayer and the biggie, Ludwig Wittgenstein. The so-called "linguistic turn" is based on the false premise that language is the *a priori* constitutive condition for experiencing the world. Wittgenstein brought this orientation to its fullest—or emptiest—expression. In his *Tractatus*, he said very plainly that philosophy cannot itself say anything; it can only demonstrate what can be said. "It can in the end only describe it [the use of language].... It leaves everything as it is."[10] How better to guarantee in advance the innocuousness, the pointless irrelevance of philosophy?

The idea that in philosophy we discover only what we already know is not confined to philosophers of language. Jerome A. Miller notes that this idea is built into philosophy's

9   Robyn Ferrell, *Genres of Philosophy* (Burlington, VT: Ashgate, 2002), p. 120.

10   Ludwig Wittgenstein, *Wittgenstein's Philosophical Investigations* (Albany, NY: SUNY Press, 1999), #124.

self-image; one "can find it in Plato, Kant, Hegel, Wittgenstein, Heidegger, and many other philosophers."[11]

Immanuel Kant insisted that any and all empirical elements be excluded from philosophy. Likewise, Martin Heidegger never made the slightest effort to address the social or historical roots of the ontological or essential tradition he attacked. Nonetheless, a few, starting with Helmuth Plessner, have advocated a philosophical anthropology, a collaboration between the two fields.[12] Whether this approach provides necessary grounding to philosophy is doubtful so far, as it implies a counter-current to abstraction, philosophy's fundamental orientation.

The rarefied thickets of philosophy are a bad joke given our painful present, not forgetting that through the ages this Throne of Thinking has ignored the fact of vast misery. As Adorno had it, "Perennial suffering has as much right to expression as a tortured man has to scream."[13] For Adorno and Horkheimer, the Holocaust was prefigured in Kant's disembodied rationality.[14]

In my opinion, Nietzsche's most valuable insight bears on this very avoidance. He saw philosophy as a false and cowardly evasion, because it has no interest in the cruelty and unhappiness in the world.[15] Especially in *On the Genealogy of Morals*, Nietzsche underscores the historical constructedness

---

11   Jerome A. Miller, *In the Throe of Wonder* (Albany, NY: SUNY Press, 1992), p. 1.

12   See, for example, Kevin M. Cahill et al., eds., *Finite but Unbounded: New Approaches in Philosophical Anthropology* (Berlin: DeGruyter, 2017).

13   Adorno, *op. cit.*, p. 362.

14   Theodor Adorno and Max Horkheimer, *Dialectic of Enlightenment* (New York: Verso, 1979), pp. 81–119.

15   Cynthia Halpern, *Suffering, Politics, Power* (Albany, NY: SUNY Press, 2002), p. 180.

of philosophical concepts. He saw Kant as a weak and watery sun, as that "pale, northern, Konigsbergian."[16]

François Laruelle's leftist "non-philosophy"[17] promises to "liberate philosophy from itself," but provides only vague rhetoric, e.g., "Non-philosophy is thought made into an ultimatum."[18] The postmodern "weak thought" of Gianni Vattimo[19] at least has the good grace of abject surrender. He explicitly accepts that philosophy can achieve pretty much nothing. Sal Restivo offers a requiem for philosophy, finding it ever more irrelevant and outdated.[20]

"Philosophy is really homesickness, an urge to be at home everywhere," according to Novalis,[21] but this urge misses its mark completely. Adorno's reversal of Hegel comes to mind: not "the true is the whole," but "the whole is the false."[22] The only whole is that of nature, its sheer thereness intolerable to civilization. Civilization at its core is a dead thing, but it continues to metastasize, along with a contagious nihilism.

Despite the presence of the very few dissenters (e.g., Diogenes, the Cynics), philosophy is part of the anti-life current. I've been called a philosopher; as I hope I've made it clear in these few words, that is a misnomer.

---

16   Friedrich Nietzsche, *Twilight of the Idols and The Anti-Christ* (New York: Penguin, 1990), p. 50.

17   François Laruelle, *Philosophies of Difference: A Critical Introduction to Non-Philosophy* (New York: Continuum, 2010).

18   *Ibid.*, pp. xv, 149.

19   Gianni Vattimo, *Weak Thought* (Albany, NY: SUNY Press, 2012).

20   Sal Restivo, *Sociology, Science, and the End of Philosophy* (New York: Palgrave Macmillan, 2017).

21   Quoted in Martin Heidegger, *The Fundamental Concepts of Metaphysics* (Bloomington, IN: Indiana University Press, 1995), p. 5.

22   G.W.T. Hegel, *Philosophy of Right* (Kingston, Ontario: Queens University of Canada Press, 1896), p. xxvii; Theodor Adorno, *Minima Moralia: Reflections from Damaged Life* (New York: Verso, 2005), #29.

MY PROBLEM WITH ADORNO

Over the years, I think that I've cited no one more often than Theodor Adorno. But I feel the need to point out a basic disagreement or two that I have with his orientation.

With Max Horkheimer, his *Dialectic of Enlightenment* collection, especially in its first, eponymous essay, points to the repressive essence of civilization. He refers to the encounter of Odysseus/Ulysses with the Sirens in Book XII of *The Odyssey*. Our hero, tempted by the Sirens' famous song, has the ears of his crew stopped up with wax and himself tied to the ship's mast in order to resist their charms. The sirens represent eros and freedom, and Odysseus must refuse them and sail on toward civilization and domination. Adorno and Horkheimer thus recognize the fateful, even fatal choice that is made.

But at the end of the essay, they introduce a caveat that undoes its radical insight. Ignorant of what is now orthodox anthropology, they assert that domination of nature/civilization is, after all, required for society to survive the anti-social power of nature. A profound misjudgment, given the fact that *Homo* species lived in harmony with the natural world for many thousands of generations, without domination/ domestication/civilization.

In *Dialectic of Enlightenment* and elsewhere, Adorno argued that there is no enlightenment without myth, no progress without regression. His provocative work, however, cannot hide the fact that such formulations are a kind of fence-straddling, that he is, at base, a progressive who accepts modernity and its progress as necessary.

"The Unleashing of Productive Forces" section of his *Negative Dialectics* refers to the word "unleashing" and its overtones of menace, while reminding us that there are certainly "periods when the living need the progress of productive forces" (p. 306). In fact, nowhere does he indict productionism itself.

He turns on his beloved Schoenberg in just this context, in reference to Schoenberg's *Die Glückliche Hand*, which as a critique of reification he labels reactionary (*Philosophy of Music*, p. 39). For it is aimed at the division of labor and not, properly, against the social relations of production. That is, Adorno's focus is not on the fundamental maiming of the individual—and society—by division of labor, but on who is in charge of production, who oversees division of labor.

Also worth noting: Adorno seems to agree with the Stalinists' official condemnation of Georg Lukacs on charges of idealism in his treatment of reification (*Negative Dialectics*, p. 190). It is apparently idealist to attack reification too fundamentally, reminiscent of Hegel's counsel that some alienation is necessary in service to World Spirit.

Adorno's critique of Heidegger, in *Negative Dialectics*, is complete in many respects, but also includes what seems to be a blanket ban on originary thinking. Heidegger approaches origins, the primordial roots of Being, but his quest, lacking any particulars or evidence, recedes ever further into the past. His approach is thus without context or content, as Adorno correctly sees; the examination of lost immediacy should nevertheless not be foresworn. Adorno's judgment that "There is no other way to break out of history than regression" betrays his progressivism. Walter Benjamin's last work, *Theses on History*, took an opposite conclusion: the only radically utopian project is precisely to break out of history's continuum, to thus enact the radical break with the ruling order.

Lacan, Derrida and others have sought to bury us in history, to banish recovery efforts. They are married to what has prevailed so far at a primary level. In this respect there is a similarity with Adorno.

Theodor Adorno has made huge contributions to social theory and has been in many ways a model original thinker to me. The above remarks in no way seek to comprehensively confront Adorno's breadth and depth.

"Have you ever been experienced?"
—Jimi Hendrix

EXPERIENCE

The word comes from the Latin *expiriri*, to try or try out. Thus it is an act, involving both connection and intent. Also, from the Latin *peritus*, signifying peril or risk.

Experience covers so much that it is hard to determine what it excludes. To be living is to be experiencing. Hans-Georg Gadamer writes, "However paradoxical it may seem, the concept of experience seems to me one of the most unelucidated concepts we have."[1] At the same time, to approach it as a concept is to already introduce difficulties.

Experience is an elusive master notion or category, and what it signifies has varied through time. Until recently, for example, it had to do with a direct relationship. It meant direct experience, but this has shifted significantly.

There was also the strong connotation of experience as unique, original. Walter Benjamin mourned the erosion of the aura, the signature essence of what was original. Jean Baudrillard pushed this further, detailing an autopsy of any original reality. All tends toward mediation and equivalence in mass society, away from an originary source.

"What has got lost," according to David Malouf, "is much that once belonged to our direct sensory experience."[2] He referred to a new kind of distancing, recalling what other commentators, from Theodor Adorno to Frederic Jameson, have observed as to the waning of experience and affect.

The word is deeply coded, however. Late modernity values experience for its own sake; one hears this a lot,

---

1   Hans-Georg Gadamer, *Truth and Method* (London: Continuum, 2004), p. 341.

2   David Malouf, *On Experience* (Melbourne: Melbourne University Press, 2008), p. 33.

perhaps especially in the United States. Again, its meaning has changed. More a commodified facsimile of what used to be meant by experience. Part of the pathology of modern life is its paucity of immediacy and connectedness. We are well into the era of the Experience Industry, of manufactured sensations, of proliferating Disneyland-like offerings for many who require ever-greater stimulation and distraction.

We experience an assaultive, fragmenting and numbing quality of life, and are left with not much of a sense of what authentic experience would consist of. Accomplice to the devastation is postmodern culture, which has worked hard to undermine any such concept as authenticity.

It is (and maybe always has been) mysterious how our experiences reveal the way things are. How it works. Hume, Kant, and many others have tried to address this. How do we get beyond the immediate data of experience? Again, we run into problematic terms: can the content of our experience be legitimately referred to as "data"?

There is always a need to interpret our experiences. Meanwhile, we can note what is experienced and what cannot be experienced in a particular culture, given that some senses have been sidelined by modernity. Wilhelm Dilthey found that "the present is the filling of a moment of time with reality: it is experience."[3] That is, lived experience, and the challenge is to not let life pass us by without fully experiencing it.

Returning to Gadamer: "...the concept of experience seems to me one of the most obscure we have."[4] In the introduction to his novel *Experience*, published in 2000, Martin Amis provides a surprisingly unthinking take on the

---

3   Wilhelm Dilthey, quoted in Susan A. Crane, "Of Photographs, Puns, and Presence," in Ranjan Ghosh and Ethan Kleinberg, eds., *Presence: Philosophy, History, and Cultural Theory for the Twenty-First Century* (Ithaca, NY: Cornell University Press, 2013), p. 4.

4   Gadamer, *op. cit.*, p. 346.

subject: "Nothing, for now, can compete with experience—so unanswerably authentic, and so liberally and democratically dispensed."[5] Some decades earlier, R.D. Laing correctly noted the dominant order's need for false consciousness: "It is not enough to destroy one's own and other people's experience."[6] Amis seems to have missed this truth entirely. Experience is the only evidence, but it can—and must—be deformed and manipulated.

We must uncover, reclaim the immediacy of lived experience, which at base is indivisible consciousness. To refer to experience as experience of something is itself too great a separation, as Dilthey saw it.[7] Raymond Tallis conflates consciousness and experience in a similar way and sees both as wonderful mysteries.[8]

Direct, non-conceptualized experience has beckoned to many as a goal or object of comprehension. More than a century ago, Edmund Husserl originated phenomenology, based on active experience, with the goal of its being the philosophy of experience. Husserl's lack of success is largely owing to the philosophic nature of his project. It's more than unlikely that philosophy, that most general and abstract mode, would break through, as Husserl put it, "to the things themselves."

William James, among others, stressed the centrality of bodily sensations in organizing our own experience, and also for our basic sense of self. We experience others most basically in face-to-face encounters, being present to each other. The rest tends toward simulation and representation.

---

5   Martin Amis, *Experience* (New York: Hyperion, 2000), p. 6.

6   R.D. Laing, *The Politics of Experience* (New York: Pantheon Books, 1967), p. 35.

7   Gadamer, *op. cit.*, p 223.

8   Raymond Tallis, *The King of Infinite Space: A Portrait of Your Head* (New Haven, CT: Yale University Press, 2008).

The background of this human condition is our extremely long tenure as hunter-gatherers in intimate community.[9] Four-Legged Human touches on our heritage of non-domesticated autonomy in his stunning "Wolf Encounters" essay: "Every experience I have while actually attempting to live in wildness, especially my experiences in shared space and relation with wild others, make this core reality continuously and increasingly clear."[10]

Michel Serres sees the body as the model and originator of all that we know. "There is nothing in knowledge that has not been first in the entire body."[11] He offers nothing less than a redemptive view of the relation between the active body and knowing. "Go, run, faith will come to you, the body will sort things out."[12]

How is modernity experienced at the site of the body? Raoul Vaneigem and Walter Benjamin noted the free character of children who, for a time, are permitted to explore and learn through tactile cognition. The absence of mediation doesn't last, needless to say.

At the beginning of *The Gay Science*, Nietzsche contends that "the history of philosophy up to now has been the history of the repression of the body."[13] The life-world as it is concretely experienced has yet to be taken as primary, as not formed by conceptuality. But what is given in experience remains. Eugene T. Gendlin refers to its promise: "We carry our life with us. Our bodies can total up years of all kinds

9   See Sarah Blaffer Hrdy, *Mothers and Others* (Cambridge, MA: Harvard University Press, 2009) re: human connectedness rooted in our communal hunter-gatherer past.

10   Four-Legged Human, "Wolf Encounters: Some Field Notes on Being Present, Inter-Species Bonding, and Anti-Domestication," *Wild Resistance* No. 6, p. 53.

11   Michel Serres, *Variations on the Body* (Minneapolis, MN: Univocal, 2011), p. 70.

12   *Ibid.*, p. 78.

13   Friedrich Nietzsche, *The Gay Science* (New York: Vintage, 1974), p. 32.

of experience and at any moment give us something new, a new, more intricate step."[14]

Our bodies are now explained by machines—the technological reduction of everything to empty exactitude. Virtual reality, in an unreal world, is just a part of how impoverished our life-world is. The Machine mines our experiences, monetizes them, and sells the denatured product back to us.

Beginning with the Industrial Revolution in the 18th century, increased velocity in all areas of human experience is a primary reality. A sense of being rushed, pressed for time, is a trademark of modernity, more and more. This temporal speed-up is a function of technological acceleration, and produces an increase in the decay rates of the reliability or authenticity of experiences. It's all part and parcel of the quickening tech juggernaut whose most pernicious effects (aside from rampant pollution) are de-individuation and a decomposition of the social fabric.

Just when the depletion and displacement of experience could hardly be more manifest, when postmodernity has insisted that all of experience is a culturally constructed product, other perspectives are making themselves heard. Soren Kierkegaard long ago rejected the systems of philosophy, especially Hegel's, on the grounds that life cannot be captured within any conceptual system. Much more recently, early postmodernist Jean-François Lyotard also came to realize that experience cannot be exhausted by concepts. Maurice Merleau-Ponty rejected representational explanations of bodily experience in a similar vein.[15]

---

14   Eugene T. Gendlin, "The Wider Role of Bodily Sense in Thought and Language," in Maxine Sheets-Johnstone, ed., *Giving the Body its Due* (Albany, NY: SUNY Press, 1992), p. 192.

15   Especially his last work. See Douglas Beck Low, *Merleau-Ponty's Last Vision: A Proposal for the Completion of the Visible and the Invisible* (Evanston, IL: Northwestern University Press, 2000).

The Derridean enshrinement of symbolic, representational culture, the assertion that there is nothing outside the text, is past its sell-by date. It is giving way to healthy, embodied modes, such as N. Katherine Hayles' insight "that a highly sensitive and interactive realm of experience exists that precedes linguistic expression and legitimately counts as cognition."[16]

We are programmed away from sensorial richness through the acquisition of language. Written language especially, independent of human bodies, effaces, disappears the body. For Nietzsche language was a crutch, a sign of debility.[17] Michel Serres, much like David Abram, contended that the development of language worked against our initial sensuous perception of the world. "Once words came to dominate flesh and matter, which were previously innocent," he averred, "all we have left is to dream.... If a revolt is to come, it will have to come from the five senses."[18]

Culture is being pacified, as surely as nature. In postmodernity, experience is already a virtual reality wherein emojis and emoticons replace actual communication among human subjects. Passion was once essential to the search for knowing. It was direct, and not without risk, an adventure. It is not enough to assert that the ultimate presupposition for knowledge is experience, when that item has become a poor substitute for its reality. Pain and sorrow, too, are part of the quest, but the life-world has lost so much of its affective range. Suffering itself becomes aimless.

And yet, even now, the quest for what is authentic remains, and Serres may well be right about its reward: "You'll recognize thought without fail by the health it gives."[19]

16   N. Katherine Hayles, "Foreword," in Mark Hansen, *Embodying Technesis: Technology Beyond Writing* (Ann Arbor, MI: University of Michigan Press, 2000), p. vii.

17   H. Peter Steeves, *The Things Themselves* (Albany, NY: SUNY Press, 2006), p. 12.

18   Michel Serres, *Angels: A Modern Myth* (New York: Flammarion, 1995), p. 71.

19   Serres, *op. cit.*, 2011, p. 110.

"O lost, and by the wind grieved, ghost, come back again."

—Thomas Wolfe, *Look Homeward, Angel*

O LOST...?

The temper of the times is that of emptiness, exhaustion. A sense of a disintegrating end-time in which the endgame plays itself out slowly, corrosively. Both physically and spiritually, the life-world seems to be buckling under a massive weight or force of estrangement. Dialogue, relationships, everything in peril.

As usual, this didn't arrive overnight. I think of the unrelieved negativity of the late-'70s band Joy Division. Its lead singer, Ian Curtis, was a suicide at age 23. Named for a Nazi sex-slave brothel, Joy Division's bleakness foreshadowed today's general immiseration. See Mark Fisher's *Ghosts of My Life* or his lecture "The Slow Cancellation of the Future."

As society darkens, unlikely voices make themselves heard. Beginning in the '90s, novelist Michel Houellebecq has pointed to a landscape with finally no energy left, a terminal emptiness. The overall theme of misery in his work has an eschatological quality; it's not surprising that Houellebecq, while seen as a nasty, wholly non-PC character, views religion as society's only hope. His outlook seems to be widely shared; Ben Jeffery calls it "depressive realism."

Sociologist-philosopher Bruno Latour, with his post-secularity notion, appears to think in a similar vein. Not calling for religious wars, obviously, but for an acknowledgment of belief as perhaps a last footing for deteriorating existence.

Surrender of one kind or another calls. One-time Dark Mountain nature activist and civilization critic Paul Kingsnorth threw in the towel not long ago. His reward was a *New York Times Magazine* feature (April 29, 2014) in which he announced that the idea of overcoming this nightmare is a delusion. Kingsnorth deepened the level of defeat in a June 21, 2019 *Emergence Magazine* podcast, "The Language of the Master." He argues that modern written language is a tool of ecocide, but that there is no solution. In fact, the primary fallacy of language, he avers, is in setting up reality in terms of "problems" to be "solved." Thus not only is there no hope of altering reality, the very idea or conception of doing so is delusional! Could there be a more supine response to the reigning Horror Show?

More important viscerally in our everyday lives is the amount of suffering we experience. For about a year, I've tried to be there for a very close friend who lives with suicide on a very real daily basis. I have a feeling that most of you know, or know of more individuals than ever who are in extremis.

For some, surrender is not an option. Ultimate redemption still beckons, and they will play it out to the end. As did courageous Will Van Spronsen, who died in battle against unfreedom, summer 2019.

VALUE AND ITS ENEMIES

Anti-authoritarian tendencies such as Stirnerism and nihilism are prone to aiming the "moralizing" epithet at their critics. Their goal is to undermine the "merely" ethical, so that value does not enter the picture.

They are not the only ones who employ this approach. Christopher Hemming blames an emphasis on value for largely causing the eclipse of Marxism. The focus of his *Philosophy After Marx* is revealed in the subtitle: *100 Years of Misreadings and the Normative Turn in Political Philosophy.*[1] Henning blasts this "normative turn" as little more than a regrettable "moral sermonizing."

Such reactions are as good a way as any to introduce the topic of value in general. Somewhere in his *Minima Moralia*, Theodor Adorno tells us, "there is no right life in the wrong one." And yet, of course, we seek to live at least less wrongly. We live so as to realize a life optimally valuable, a life whose values are most realized.

Life is a horizon-forming project wherein we seek to find out what matters. What part of life is unrelated to values? To deny this, in the name of whatever ideology, is to deny the fact of our evaluative life-world. Milton Rokeach, in a too-narrow understatement, holds that the "concept of values... is the core concept across all the social sciences."[2]

Kant rejected nature as a source of value, but for many of us the unbuilt world is a touchstone. Placelessness is currently succeeding, but a bond with the natural world remains a vital value.

---

1   Christopher Henning, *Philosophy After Marx: 100 Years of Misreadings and the Normative Turn in Political Philosophy* (Chicago: Haymarket Books, 2015).

2   Milton Rokeach, *The Nature of Human Values* (New York: The Free Press, 1973), p. ix.

Value is what validates, and to forget value is to accept the dominant evaluative norm or authority. The disintegration of cultural meaning and connectedness collides with a will to meaning and value. Modern meaninglessness versus the creation of meaning and value. But as Lorenzo Simpson has argued, "Meaning cannot be engineered." He goes on to ask, "Must meaning then not be, in some sense, found and not made?"[3] Life outside of domestication/civilization was based upon just such a premise: based on what nature gave, not the engineering of nature.

There is no necessary relation between meaning and value. I may know what something means but place no value on it, but even in this sense there is an evaluative aspect. It is also clear that feelings of meaninglessness and powerlessness are mutually reinforcing.

As with modes of technology, there are no lines of thought that are neutral or value-free. Every form of living has its basic or dominant values. For hunter-gatherer life the cardinal values, according to consensus anthropology, are egalitarianism and sharing. I am reminded of Walker and Shipman's meditation on the remains of a *Homo erectus* woman who lived 1.7 million years ago. In life she was afflicted with advanced hypervitaminosis A, completely unable to provide for herself. "Someone else took care of her," was the stunning conclusion.[4] Someone, probably many, valued her life in the ultimate manner.

For Friedrich Nietzsche, values are creations capable of raising us up. We are the endangered creators of all values. Max Scheler, often referred to as the Catholic Nietzsche, saw

---

3   Lorenzo Simpson, *Time, Technology, and the Conversations of Modernity* (New York: Routledge, 1995), p. 165.

4   Alan Walker and Pat Shipman, *The Wisdom of the Bones: In Search of Human Origins* (New York: Alfred A. Knopf, 1996), p. 165.

values as coming first, preceding all representational acts.[5] He argued that they are the illuminations that motivate striving.[6]

Martin Heidegger did not give pride of place to values. He went so far as to assert, in his "Letter on Humanism," that thinking in values is the greatest blasphemy imaginable against Being.[7] To bring in values is to dominate. Heidegger held that "values-thinking" is inherently "technological" in its domineering and exploitative nature.[8] It is not necessarily anything of the sort, and Heidegger's aversion to values cannot be seen as unrelated to his fealty to Nazism.

Another approach to denying value is the egoist-individualist focus on the sovereign Self. To base evaluations on one's subjectivity is an effort to take value out of the picture, to retreat to a supposedly anti-ideological solipsism. Of course, as Alfred Adler reminds us, "No isolated persons are to be found in the whole history of humanity."[9] Thus the egoist enthroning of the self is an illusory retreat. In this misguided vein, Georges Bataille pushed the virtue of human isolation: we are essentially solitary creatures who only join with others out of weakness and lack of courage.[10] Ethics philosopher Emmanuel Levinas saw rather the opposite: sensibility itself as a valuing orientation to the Other. As did

---

5   Philip J. Harold, *Prophetic Politics* (Athens, OH: Ohio University Press, 2009), p. 46.

6   *Ibid.*, p. 37.

7   Quoted in Steven Hodge, *Martin Heidegger: Challenge to Education* (New York: Springer, 2015), p. 25.

8   Simon P. James, *The Presence of Nature* (New York: Palgrave, 2009), p. 12.

9   Alfred Adler, *Social Interest: A Challenge to Mankind* (London: Faber and Faber Ltd., 1938), p. 282.

10   Tzvetan Todorov, *Life in Common: An Essay in General Anthropology* (Lincoln, NE: University of Nebraska Press, 2001), pp. 33, 37.

his mentor Edmund Husserl, who found in every type of consciousness "the lived experience of value."[11]

We live in a post-truth, post-meaning postmodern era, in which value is almost as big a casualty as meaning. Despair mounts, and probably the biggest danger of postmodernist culture is, as Simpson has captured so well, its basis in technology. We are in grave danger of being completely dominated and domesticated by technology, the wellspring of a world lacking in both meaning and value.

Value is plural, various in its applications, which Kant did not see. Nor is it a substitute for analysis. But life is irreducibly normative and this is certainly no time to exorcise value. Our values should not be hidden, but clearly marked. What would be the value of doing otherwise?

---

11 Christian Lotz, *From Affectivity to Subjectivity* (New York: Palgrave, 2007), p. 57. Many thanks to Nathan June for very helpful exchanges.

WHERE ARE WE?
WHERE ARE WE GOING?

In closing, let's get into the weeds
a bit. Three of us anarchists explore
issues, approaches, ways to go
forward. The interview is an example
of dialogue that tries to sort it all out.
Or at least some of it!

WHERE ARE WE? WHERE ARE WE GOING?
A Conversation with Steve Kirk and Bellamy Fitzpatrick

STEVE KIRK: I guess let's just jump right into it. We kind of started this conversation around the idea of the anarchist milieu and where that was and given that John has Anarchy Radio where he's able to kind of talk about these things on a week-to-week basis, and Bellamy coming from Free Radical Radio way back, I think it would be really interesting to hear just general takes on where we are as anarchists?

JOHN ZERZAN: Would you like to go first, Bellamy, on that?

BELLAMY FITZPATRICK: I'm from Ohio so I think it would be almost morally impossible for me to go first in a situation like this so you can go ahead.

JOHN ZERZAN: Okay. Well, the milieu, I think you're right, Steve, to throw in Anarchy Radio because I kick this around—more or less as a constant I suppose. I would say that if we're under the sign of nihilism and egoism the so-called milieu has just about nothing to offer. In large part it doesn't try to offer anything because part of the—part of the thing there is, there's such a widespread feeling that nothing can be done, there can be no transformation, no prospects at all for overturning the nightmare. So, what follows from that is going to be kind of limited—going to be kind of uninteresting, at least to me. And if we think about being anti-authoritarian, or anarchists, to me that includes, primarily, what do we have to offer. What are we sharing with people? How are we helping people understand things and conceivably go forward? If you don't have that thing in mind then, well, where are you at?

And I think in general, and I guess I dote on this maybe all too much, this is a postmodern culture. And, right away, that's so disarming or debilitating in my view. In other words, for example, if you don't believe there's any stable truth or meaning, and maybe not any outside, actual world, then—again, you're kind of crippled. You kind of go nowhere because you don't believe you can go anywhere or should go anywhere. Or the—and this maybe—I hope this is a little out of date, I have some idea that it is in terms of the postmodern culture—the idea of overview. The idea that the totality is totalitarian. To try and get a grasp of what's going on in general is bad, you don't do that. You throw that out.

So, I think that's coming back. I was just reading today in *The World Interior of Capital* by Peter Sloterdijk he starts right out, the first page he's saying, I have a grand narrative, I have an overview and I'm going to give it to you. I haven't gotten into it that much but he's not shrinking from that, he hasn't been scared away by the—by Lyotard and all the others who announced the very end of metanarratives or overview.

So, anyway, in general, in terms of the milieu, I just think it's in sorry shape and hardly makes an effort.

STEVE KIRK: Bellamy, would you like to talk about that a little bit.

BELLAMY FITZPATRICK: Sure, so I agree with a good deal of what John said. I would say that there is a certain tendency towards defeatism and that that is harmful and there's no need to really to sign on to that philosophically. But, going alongside it is a different tendency that I think we see represented by those who are more open to engaging in kind of big-tent activism and that that carries its own set of problems that are separate from the ones John outlined.

And I should bracket all of this by saying, you know, I live rurally now. I obviously used to live in the Bay Area and was very much in and of the subculture. But I didn't come to anarchism through the subculture or the milieu, I came to it through philosophy and I—in certain ways I kind of have one foot outside of the subculture now. So, my comments are indexed to what I saw when I was in the Bay Area, what I've seen in New York City when I'm there, my time that I've traveled to different urban and rural locations with anarchist activity over the past several years. And, of course, what I see through media whether in print or through the internet. So that's my vantage point, that's where these comments are coming from.

When I talk about this big-tent activism, something I've been disappointed by, predictably, is the way that, in the past few years, anarchists seem to have been very much led by the mass media. And part of that is this big reaction to Trump, as if he's some kind of paradigm shift, which I really think he's not in any significant way. And you see that anarchists have, for instance, taken up this immigration/deportation issue, it's become this major hot point of anarchist activism even though deportation rates were higher under Obama than they have been under Trump, even allowing for the fact that Trump hasn't had as much time. The activism around the detention centers, these detention centers have existed for a long time, they pre-dated Trump. And yet you see people acting as if from this place that everything has changed now and, therefore, our priorities need to be changed. And that's just disheartening because it tells me that there's a lack of sober analysis and that anarchists are, in North America, being very easily led by current events, the sort of topic of the moment.

And if someone wants to say, no, I'm doing this for prin-cipled reasons and it doesn't have to do with Trump, that's fine, good for you. I'm not seeing that these are necessarily bad struggles to be engaged in. What I'm concerned by is what seems to be the proximate cause for them. Which is this being very led by the media narrative, being very led by this 24-hour news cycle, which I think—it's easy to get sucked in by if you don't really step back from that.

And, of course, going along with that is the way that—I was talking to someone last week from the, this Inhabit project. Capital I. And he was saying it's as if anarchism in North America has become Antifa, so much of the energy has been taken up by that. And, in some way, it seems like something where people are very led by the media that the alt-right is the thing we should be so concerned with and they get lots of media attention, therefore they become a hot point for anarchists, even though really their power is very limited. And it becomes vanishingly small when you consider it in comparison to just the everyday commodity life, the assault on the living world, the meaninglessness that John was referring to—the kind of disbelief in any kind of truth or meaning or value.

These issues to me are far bigger. And if anarchists are going to be significant in this world, I think they need to really have principles and an analysis that they can come back to and act from—that doesn't just get swept aside by whatever the new issue of the 15 minutes is. And I feel like I'm going on too long, so I'll just cut it there.

STEVE KIRK: No, I think it's fine. I just thought it was interesting that you kind of immediately went to the leftist anarchism—and I think for good reason, it's obvious that there's quite a bit of momentum there, even if it is tied to this 24-hour news cycle, I think you're on to something there. But it also occurs to me that that is also somewhat cyclical of Republican administrations and sort of leftist campaigning in general. It seems pretty perennial. It's something we should be used to by now. And it is somewhat disheartening to see some anarchists take on this more traditional, RCP-type tactic of straight-up co-opting causes and trying to use it for propaganda. I guess there's a door there but I'm not sure.

But, John, I thought it was interesting that you didn't mention that? Did you want to comment there?

JOHN ZERZAN: Well I couldn't agree with Bellamy more, I think this whole Antifa thing is lowest-common-denominator leftism and in order to rule out everything else they have to inflate the global neo-fascist wave that's sweeping over everything.

Whereas, it is very minimal I think, I mean that could change I guess, but I don't see—plus, as you both were saying I think, if there wasn't any neo-Nazi or neo-fascist thing how would anything else change? I mean I'm not brushing under the carpet the whole thing about racism or anti-Antisemitism and other forms of bigotry but if you make that your whole world you are crowding everything out. You're blocking out all these other questions.

You know, I would like to say something, Bellamy referred to this in passing, this being led by the media. I guess, to some extent, I'm led by the media in the sense that—and you can't conflate the two—but I would say I'm led by reality, such as the mass shootings that are happening every day. And just an hour ago I was searching the web a little bit here: Suicide rate up 33 percent over the past 20 years, since 1999. I mean, these are things, and I got it from the media I guess is what I'm saying. But there's all too many anarchists, they've got their analysis of their position, but they don't seem to have a clue about the outside world, they don't even know. I hardly ever see it referred to in what we—I guess the more well-known anarchist sites or what have you. I don't even see it.

So, what relevance do we have if we don't even know what the fuck is going on? I think the enormity of this should make us even more willing, or ardent even, to try to say, well, what do we have to offer? Do we have any inspiration or analysis or anything? Or are we just going to—I'm a nihilist, I'm an egoist. Who the fuck cares?

BELLAMY FITZPATRICK: Right, I hope it would be clear from what I said, I'm not suggesting that people stop paying attention to the events of the world...

JOHN ZERZAN: Oh, no.

BELLAMY FITZPATRICK: ... I'm suggesting that they be careful and realize that the framing, what gets talked about, how things are...

STEVE KIRK: How they're placed in the media?

BELLAMY FITZPATRICK: Yes.

JOHN ZERZAN: Sure, yes, understood.

STEVE KIRK: I think that's a really interesting point to come back to. Something you both touched on is this—sort of the obviousness of the crisis on a deep interpersonal and community level that seems to be kind of in front of everybody's faces—the suicide statistics, the mass shootings. And the denial of that situation, to me, seems no better highlighted than in the coverage of the mass shootings where they become 100 percent co-opted and you can see the media narrative is taking off on its own. Have you both felt that way?

JOHN ZERZAN: Yeah, that's why we need to step in and—what is society? What kind of society is it that you get every pathology now—that you might have not even dreamed of just a few years ago and they're just all over the map. I mean, do we have something to offer there? We better. I think we do. It goes down to fundamental things in terms of mass society, domestication, all the rest of that. You don't hear very much of that in the milieu either although I think it's making its way, some I would say deeper critique that's fueled by something. Fueled by the massive negativity for one thing. Things are just getting desperate. On every single level. I see friends just being chewed up by it.

I've been, for months now, I've been spending a lot of time with a very close friend who's very suicidal. And I just see more people—that's anecdotal obviously—but I just think it's getting so scary. It's just—are we doing something? Or are we just in our little niche or our little clique. Or, anyway...

BELLAMY FITZPATRICK: I guess, I essentially agree with what John said but I guess I would—at this moment maybe it's interesting to ask, if you were to try to give people who are hearing that message a few points of action I'm curious what those would be for you, John?

JOHN ZERZAN: Of course, my focus is writing, public questioning, the critique.

BELLAMY FITZPATRICK: Sure.

JOHN ZERZAN: And how to try to do that while being aware of the traps of the media and the spectacle, in a context where we're not given much room or platform to connect with people. And you know sometimes that takes the action and if nothing is happening of course it's all the harder to be heard. Just one example, some group, ALF group for example, burns down a McDonald's and they have a communique explaining why—you might get to read that communique because they burned down a McDonald's and it was in the news. Again, it's a media thing I suppose.

But I don't know. I mean, we don't know what to do, I don't know what to do. I'd like to have a riot every day or something (laughter) but it's usually not on offer, it hasn't been on offer for a while now so that's tough.

BELLAMY FITZPATRICK: Sure.

STEVE KIRK: What—I guess, in the same vein, would you have something that you would propose, you know to present to somebody?

BELLAMY FITZPATRICK: Yeah, I guess this is sort of apropos our earlier topic. Lately I've been thinking about the—trying to reflect critically on the media efforts I've done and what's felt successful and what hasn't and I've come around to thinking that the milieu is—I mean, this is a perennial criticism like the ones I raised earlier—but that it's small and it's fairly insular. And I've been toying with the idea of trying to reach a broader audience and thinking about what—rather than taking it to what John was saying, the kind of lowest common denominator left-wing issues, but rather as a way to appeal to discontent that might be felt across the political compass or however you define it. And what...

JOHN ZERZAN: I wonder if I could ask you, Bellamy, vis-à-vis Backwoods—I'm really impressed by it and I have a feeling that that is—are you getting good connections from that? Is something going on with that?

BELLAMY FITZPATRICK: Yeah, I mean I would say it's reaching a different audience than the podcast projects I've done before and I think that is a positive step. But I've actually been thinking, okay, what's an even kind of broader point of tension that doesn't end up shallow. And I've come to feeling that the point of tension right now, globally, are related to the technology issue, the mass society issue—understood as the death of community, the death of feeling tied to any group of people, any particular place—and globalization.

And so I've lately come to think that if anarchists, understood in the really robust sense of anarchists that I think the three of us share, that those should be our major points and that, perhaps, then, the points of action would be radical decentralization, pan-succession, Luddism of some form and that perhaps those could have purchase around the political compass. Not just anarchists but people on various parts of the green spectrum, maybe some form of small communitarian socialists, certain types of conservatives and libertarians, people with certain traditional religious beliefs and I think that maybe there could be a kind of meeting point of the myriad marginal tendencies around those issues. If we agree that, yes, global society is bad, high tech is harming us, we all want community—and maybe we totally disagree about what that looks like—what our ideal would be—but we agree that this is not appropriate and if we could form smaller, insular communities that that might be a common starting point.

And I realize that might sound heretical to certain types of anarchists who say, oh how could you countenance some of those people—and it's like, well, things are looking really, really bad and we need to have a broader appeal than what we do right now. We can't just be talking to ourselves.

STEVE KIRK: I would agree. Interestingly, you would probably have critiques from a couple different angles there. One with your inclusion of people and then the second with what sounds like possibly a program of political reform in I think—it would be interesting thing to see what the response is, and it will be interesting to see what the response is there...

BELLAMY FITZPATRICK: I'm not sure I followed your comment there...

STEVE KIRK: Well this idea of laying out a practical political strategy that could coalesce into these sorts of movements of succession or subsistence—that it would be purposeful and political across a large area or globally.

BELLAMY FITZPATRICK: To be clear, I don't mean to try and lay out a political program so much as, insofar as I'm going to participate in media and disseminate ideas, I personally feel like this is maybe the point that has some potential and it's not that I would try and bring everyone together in some kind of pan-secessionist party or something like that...

STEVE KIRK: Right.

BELLAMY FITZPATRICK: Rather, try and disseminate this sort of separatist, decentralist ideas and say like, look, we all have things that we don't like about the way things are now and what is really tying that together? It's the mass technological infrastructure. It's the idea of these global or very large societies and none of us are going to get what we want as long as those things continue.

STEVE KIRK: Yes. Thanks for the clarification. I just thought in the way that you were speaking about it, not that was necessarily that was your intention but that would be a possible critique of the way it was laid out.

JOHN ZERZAN: Good to focus on that. I totally agree, we shouldn't be afraid to be open and reach out. I mean, there are practical needs if we're going to get anywhere and that's—what Bellamy said, these are the high points, these are the key things. And so that stuff has to be present. To try and connect with people so we're not just talking to each other and as things worsen I think there's a good likelihood that people—I mean on one level I think it's kind of obvious, that people have soured. I mean you can't find anyone to defend anything, but the thing goes on because of inertia, and other reasons.

But it's not like we're going to be running up against people that are just going to rush out and, like Pinker [Steven Pinker], defend civilization and everything is wonderful. (Laughter)

It's just so ludicrous, it's so preposterous that anyone would, well, try and fail so grandly. It's just like a case in point. I mean, list all the things that are stupid in defense. You and Chomsky and these others—it won't work, it's not happening.

STEVE KIRK: Yeah, six hundred graphs don't make your civilized life any better. (Laughter)

And then the whole issue of him with Epstein too, it just seems like what a cruel revealing of this fraud.

JOHN ZERZAN: Yeah.

BELLAMY FITZPATRICK: Wait, you're saying Pinker had connections to Epstein?

STEVE KIRK: Yes.

BELLAMY FITZPATRICK: Oh, wow.

JOHN ZERZAN: Delicious. (Laughter) So, it's good that this is exposed. You can more easily connect the dots. This is what this whole racket is about—or that's part of it—the corruption itself isn't all of it. But you know the whole damn thing is rotten and it's hurting everybody. Every species.

STEVE KIRK: Yeah.

BELLAMY FITZPATRICK: Yeah.

STEVE KIRK: It kind of gets me into this thing of possibility in general. And that's something we're really interested in. Not so much in the—I guess I hadn't considered it as much, and I need to, consider it a little bit more in the way that you were framing it, Bellamy. But the way that it's sort of presented in Backwoods and I think it's a common thread through a lot of anarcho-primitivist writing, this possibility that—whether that's there's this wild human for you to become or there's these openings to have subsistence, however you want to frame. To me it feels as though there's a ton of possibility as we have this sort of fever-pitch political climate—that despite everything, despite the news coverage and all that, there's a lot of people who are engaged and are willing to listen to a lot of different things and I wondered what you thought about that.

JOHN ZERZAN: Well I've seen quite cracks of the armor, I guess, not much, but more than I used to. I got on public radio, what, about a year ago I guess, it was to talk about *A People's History of Civilization.* And I recall, like 20 years ago, there was this woman from Australia did a very fine film about forest defense and she had the cops in there, the timber guys, timber CEO and everything—it had the 'balance' that's usually very necessary (laughter). So, she took it to public broadcasting here in Oregon and, man, she practically got arrested.

STEVE KIRK: Oh, wow.

JOHN ZERZAN: They practically treated her like she was some terrorists. So, I just thought, those fucking liberals, it's going to be a cold day in hell when they're open to anything. And then, I get an invitation to have a half an hour on the talk show and I thought it was a joke at first. I just thought, yeah, sure. But the guy was totally interested.

So, I mean, maybe that doesn't mean much of anything, but I see some small signs that there will be maybe more of a chance to have a voice, to be a little more accessible. But, then again, we get into the whole question of media...

STEVE KIRK: Right.

JOHN ZERZAN: (Laughter) ... which can be tricky. I've been called a 'media slut' and I sort of agree.

(Group laughter)

JOHN ZERZAN: I agree with—people I respect, they think you should never do that, you should never talk to media. Well, I disagree.

BELLAMY FITZPATRICK: This is a safe pace for sluts right here, so. (Laughter)

JOHN ZERZAN: Yeah, that's a whole interesting topic in itself, how you navigate that, or don't you navigate that or what?

STEVE KIRK: Any thoughts on this, Bellamy, want to join in?

BELLAMY FITZPATRICK: Just thoughts on possibility broadly or...

STEVE KIRK: Or specifically as it's laid out in—I mean, I guess there's kind of these—in permaculture—and I have a lot of experience with people who are interested in permaculture out here or who are practicing it and in that experience it seems to me that they're very open to a critique of civilization but their default critique has been one that's been presented in permaculture books, which seems pretty limited and pretty anthropocentric in a lot of ways.

BELLAMY FITZPATRICK: Sure.

STEVE KIRK: I wonder though, just that vast interest, is there a possibility to pull on these strings that you were talking about earlier: mass society, the technological deprivation of literally everyone who's existing in modern life—can we pull at those strings to sort of creating something that's more vibrant and more anarchistic in the realm of what we were talking about in the beginning?

BELLAMY FITZPATRICK: Yeah, I do see some positive signs in spite of all the negativity. Those have to do with the fact there is a lot of discontent and perhaps the would-be masters of the universe have pushed too fast with some of what they want, like the high-tech phenomenon, the globalization. And you see a lot of pushback from that, again, across the political spectrum, including from some of those we may not like for various reasons—but, still, the discontent is there. I mean the confidence in public institutions in the United States is very low right now, if you get these sort of polls of— questions like: Do you feel like your representatives reflect your interests or hear your concerns? I mean, you see again and again from polls that are asking those kind of questions broadly that people feel very negatively about those sorts of things. And then these sorts of psychological issues that John often talks about are so in your face, whether it's the hikikomori phenomenon or the suicide phenomenon.

So, there is a feeling I have sometimes that—I'm not an accelerationist, I don't like those ways of thinking—but sometimes I feel like the elite are pushing so hard that there is going to be a backlash. I mean, Laura Drake is very concerned about a backlash. So that, in itself, is a good sign.

And, again, I guess, this is just me having my new idea of the moment but the idea of a certain meeting point across various malcontents on these issues of technology, mass society. I think there is some kind of possibility if we're willing to think—to really focus on this issue of radical decentralization.

And I guess, I'm going on a tangent but just bear with me, that's something else that I have a real problem with the North American anarchist milieu, is many of them seem to really want a globalized society. And they imagine that they can push these sorts of—you know, the woke morality on the entire world. And that would be stifling and impossible and totally authoritarian if it were ever to be carried out in any meaningful way. And so, I think we need to absolutely abandon the idea of a global society that's going to reflect our particular set of values. I think it's fundamentally non-anarchist and absolutely dangerous way of thinking, that's the kind of thinking that led to these awful communist revolutions that we've seen play out time and time again.

I feel like I got away from your question.

STEVE KIRK: Yes, but I think it came into an interesting one there, which is the sort of—and I don't think you mentioned it in your kind of litany of things that you were talking about that could starting points for people to become engaged with the critique, is that—this rejection of this leftist paradigm, which I think you both...

BELLAMY FITZPATRICK: Rejection of the world-society paradigm. We have to get away from the world-society paradigm.

STEVE KIRK: John, would you put that in similar terms you think?

JOHN ZERZAN: Well, yes, I think so. The emphasis on the radical decentralization I completely endorse that. So, it's a matter of particularity and getting your hands dirty. For one thing I've noticed this, as obvious as it is, if you seriously want to get rid of civilization you better be ready to live without it. The plane is going straight down but you don't jump out of the window even though you're going to be dead in a few seconds if you just sit there. I mean, I don't know, I respect the practical effort, working on these things. That's part of it too, very obviously I think...

STEVE KIRK: Yeah, we're not going to get very far if we can't figure out subsistence on a basic level.

JOHN ZERZAN: So, you know it's easy to categorically reject certain things but—face it. Here we are. In this very real situation. So, if we're talking about some kind of transition or some kind of process away from this well, if we are, then there must be actual steps and what parts would make up that possibility.

There's another emphasis, at least fairly recently, of yours, Bellamy, very intrigued by. When I read the panpsychism essay in Backwoods No. 2 and the whole question of the relationship of spirituality of the 'political'—very intriguing. And, also, and not just to throw out a whole welter of questions all at once, but one other one would be the—where is the practical connection there? We've been talking a little bit about real moves and possibilities among other people. Anyway, I was just very, very intrigued by the piece and wonder if that ties in at all with the other emphasis of trying to reach out and do something, some actual moves in the world?

BELLAMY FITZPATRICK: Yeah, I think it does. I think maybe if I were to amend the list of things I gave before of radical decentralization, critique of industrial technology and whatnot as potential meeting points for people from a wide array of political and philosophical stances I guess the other one to throw in there would be something like a return to a true spiritual way of being.

And I—the biggest shift in my views, probably that I've ever had, was probably about a year and a half to two years ago where I realized that I was just really wrong about the spiritual question and—it would take a long time to explain but suffice it to say I do think now that there's a kind of sophia perennis or core human spirituality and I think it's expressed itself in various way at various times that you can sort of—you can see the common threads among a wide variety of human expression. And this is not a new idea. I mean this is—you can see it in the capital-T Traditionalists or people like Aldous Huxley or other figures have pointed this out.

And I basically think it's true. And I think it is something that we need to come back to if we are to really be liberated. And I think our lives do have objective inherent meaning and there is a capital-T Truth and a capital-G Good and all these sorts of things that I formerly was very averse to.

And I think my reasons for being averse to it were not ill conceived, but I've had a shift. And that piece that you're referring to, John, I'm going to keep developing that and I've had some new thoughts since I wrote it and so that's going to keep being developed in future media projects of mine. But I think that is, again, an essential thing for us to be pushing that I think could allow anarchists to reach out to people who normally might be deaf to their appeals.

And I think it's a real mistake that there's this strong atheist streak—not absolute by any means—but strong in anarchism that anarchists have kind of inherited from Marxism and I think it's really just a historical accident because of the things that were going on in Europe in the 18th and 19th centuries, that basically to be a free-thinking intellectual you had to be an atheist because the church was seen so much as this instrument of authority. And that's all fine and well but there's no reason to be this hardcore atheist materialist, scientistic person. Because if you take that point of view seriously, the atheist materialists scientistic view, if you really take it on it means your life is meaningless, there's no such thing as good or beautiful or free will—and everything in the universe came out of nowhere by a complete accident, it's going nowhere and it's just a bunch of billiard balls colliding together and everything that you think your life means is just a hallucination created by your brain. And I think that's not only an awful way to feel about your life but there's no good reason to believe it philosophically.

And I think it completely dovetails with the sort of postmodern stuff that John was talking about earlier.

STEVE KIRK: Yeah, most definitely.

JOHN ZERZAN: Yeah, I remember my own sort of—well, not putting words in your mouth—but sort of awakening. I think it was in Turkey I gave a talk and afterwards this young woman said to me, I think this whole green anarchy thing is, at base, a spiritual movement. And I was just gobsmacked. Wow, I never thought of that. And then unfortunately she had to take a bus because I wanted to—I tried to ask, what do you mean by that? What's going on there?

But she had to leave. But, anyway, it opened me up more—or, opened me up period to that and I think she was thinking about—whether you're talking about wholeness or communion with the earth and that sort of thing. How is that not spiritual? How is that not connected to—and this is my word—but, you know, in terms of authenticity and the spirit.

I mean, well, it is. And I was glad to hear it. It was just an eyeopener. Okay, we don't have to be against it. We don't have to have this wooden materialism, kind of 19th-century like you were referring to, Bellamy. That doesn't go anywhere.

It doesn't appeal to anyone, especially now when our very souls, if you want to put it that way, are just being hammered. And we're in just such an attenuated place. Not to mention the suicides or the drug use and just so many countless things. Man, maybe we've never needed something like that more. Not to reify it but you know what I mean, more awareness of that. What's going on? Where are we? What are we supposed to be doing here?

BELLAMY FITZPATRICK: Right, if I can press you a little bit on that, John. I mean, when someone reads your writings it's obviously a plea for a kind of objective value, objective meaning to life, a desire for a closeness to nature. And I guess I just wonder, where does that come from? In the sense, do you want to say that nature is alive, that it's sentient in some way or that there is a transcendental dimension to things? And I realize I'm putting you on the spot so if you want to say let's table that for now—but I am curious.

JOHN ZERZAN: Well, I like that. And you know reading your piece in Backwoods made me think all the more about that. I mean, because it doesn't—I'll just throw this out, speaking of Gaia or an instance of that, you know, James Lovelock—well, the guy is in favor of geoengineering and nukes. (Laughter)

BELLAMY FITZPATRICK: Yeah, I know.

JOHN ZERZAN: So, he doesn't seem to—I mean what is he taking from that. That doesn't make sense to me. But I do think there is—and I don't mean to dismiss it because of his limitations, I don't mean that—but I mean, rather, it does make sense to me and I certainly don't reject that, that concept or that vision. It sounds pretty cool. Yeah, there's more thinking, I'd like to kind of get down with that more.

BELLAMY FITZPATRICK: Sure. Yeah, he's a real disappointment, actually; he's also a huge advocate of nuclear power.

JOHN ZERZAN: Right, right.

STEVE KIRK: And I wonder how this would normally be experienced locally. In anthropology they would normally talk about it in sort of animistic terms. But it seems on a local experience—and I'll just go off here and maybe it won't work—but there's a great book called *Soul Hunters* by Rane Willerslev in which he explores this Siberian hunting culture and how they have interacted with their local land and the animals and this very complex—there's really no species in their worldview, they're all interconnecting and interrelating, especially in a complex dream world. And I wonder how—it always occurs to me that those sort of dreams, where you're walking the land and you're exploring and coming upon an animal that you will then hunt, that that only happens when you're in a very localized world, when you're experiencing the land over and over again in a very immediate way that's also not geographically related to how we see land in civilization. So that did occur to me when I was reading "What Does the World Desire?"

BELLAMY FITZPATRICK: Yeah, I think to have any authentic spirituality you have to be spaced in a certain—I was going to say bioregion but even more local than that. And you need to have a real kinship community and—but I also don't think that acknowledging that means that we should then take this attitude of despair and say, okay, well, I'm too fucked up and I'm too broken to connect with any of that. No, the connection is possible all the time because you are at all times within the world-soul, you can't actually be disconnected from it, you can only sort of block it out and disassociate yourself from it. But that potential for connection, I think, is possible all the time. And I think it should guide us.

JOHN ZERZAN: Yeah.

STEVE KIRK: Yeah, I think that sort of brings it in to this idea—and that's also what I took a lot from that piece was this sort of hope in there. And I just wanted to read, you end the piece and I'm going to skip around between a couple of sentences but you say: "We must each live this virtuous way as best we can, always patiently moving closer to it, and speak of it as truly as we can, reaching whomever will listen." And then you close with: "Our great and terrible task is to revive it once again, first in ourselves and then in others."

BELLAMY FITZPATRICK: That's pretty good. (Laughter)

STEVE KIRK: I thought it was quite good and it had quite a lot of hope, I think, embedded in the tone of that.

BELLAMY FITZPATRICK: I suppose my views on a lot of things have changed, like I said, and I suppose I think now that it's so important to act from where we are each day and try to push toward the good and that an attitude of trying to say, oh well, what difference does it really make in the broad scheme of things or I'm just going to die is just completely the wrong way to think about things because the reality of our spiritual relationship to the world is such that that struggle, that kind of pushing against what we see as negative, is inherently meaningful and inherently positive even if it feels like a drop in a bucket. So, I just have no—I don't want to countenance at all these kinds of—this kind of defeatism. I just think that the attitude that leads to defeatism is fundamentally wielding the wrong metric in the first place.

STEVE KIRK: Great. Yeah.

JOHN ZERZAN: I agree, why should we slam the door on possibility, that doesn't make sense to me. Maybe we'll lose. Maybe we'll never get anywhere. But, man, you've got to try. And I think that's, to refer to Adorno, the need in thinking.

And not just thinking. The need. The need to be part of it and realize that we're a part of it.

STEVE KIRK: Any other topics people want to bring up?

[Editor's Note: We decide to discuss veganism]

STEVE KIRK: I don't really see where this narrative gets us. And I don't see how you can possibly draw these species lines across the world into what foods to consume and not to consumer. I mean, what are we even considering vegan; if we go back six million years, we have insect consumption. So, it seems very sticky.

JOHN ZERZAN: Yeah, but it is a strong thing. It has a strong appeal. And as much as I think Ria [Ria Montana], for example, her ideas on pre-history are just dreamed up—but I do have strong respect for her, I really do, and I don't want to...

STEVE KIRK: Yeah, I like her quite a bit.

JOHN ZERZAN: But this whole veganism thing, I guess we could spend a lot of time sorting it out, the appeal, whether it's factual or not and all the rest of it.

BELLAMY FITZPATRICK: Yeah, I have full respect for Ria, she's always engaged with me in a very good faith way, a very warm way. I appreciate all of that. I think her heart is in the right place. I guess, both of you have seen the engagement I had with Layla [Layla AbdelRahim] where I just felt like when the question of veganism got down to nuts-and-bolts there was just no defense of it, no explanation. And I don't see how a vegan could survive outside of technological infrastructure that provides for their food needs. I've thought about whether it would be possible to do it where I am, and I just come back again and again to—it would be very difficult, require close management of large areas of land and, still, even then, lead to a diet of dubious healthful value. I mean, can anyone name any traditional society or indigenous group that subsisted that way? I see that Ria tries to claim that Neanderthals did...

STEVE KIRK: Yeah, she throws out an archaeological find with low meat consumption...

BELLAMY FITZPATRICK: Okay, so low meat consumption...

STEVE KIRK: That's my understanding of the paper and I should say I have not read it. But that is my understanding of the paper as it's been characterized to me by others.

[Editor's Note: I have read it now. The study in question is: "Neanderthal behaviour, diet, and disease inferred from ancient DNA in dental calculus." The study shows high meat consumption among one group of Neanderthals and a varied fungi/plant diet with no evidence for meat consumption in the other. In other words, the cohorts of the study showed highly varied diets. Other studies of Neanderthal diet have showed cohorts of almost exclusive meat eating. This genetic dental calculus study may also provide evidence for self-medicating with plants and fungi. However, the study did not show Neanderthals were broadly vegetarians and was characterized thusly by the authors: "...our first genetic description of their diet supports evidence that Neanderthal groups across Europe used multiple subsistence strategies according to location and food availability."]

BELLAMY FITZPATRICK: Low meat consumption is not no meat consumption, right?

JOHN ZERZAN: No. You can find various people in the ethnography, the ethnology, that didn't eat a lot of meat. But what does that have to do with anything?

STEVE KIRK: Well, it's very different across the world but you get like honey-hunters in Southeast Asia probably have some of the highest non-meat or plant consumption or honey consumption—whether we want to consider honey a vegan product or not, we can table that—but then you have this huge ethnographic problem where anthropologists were categorizing all small animals as gathered because they were done by women and children. So you have this, even in the best-case scenario, and I know she's [Ria Montana] familiar with these examples because she does cite *Us, Relatives* by Bird-David who has done a lot of work in the area—but, anyway, I don't see any but maybe I'm putting my predatory blinders on.

BELLAMY FITZPATRICK: Right!

JOHN ZERZAN: I haven't seen any.

BELLAMY FITZPATRICK: I think the whole vegan phenomenon is, to me, fairly obviously derivative of modern liberal values and trying to extend the human-rights paradigm to non-human animals and it's just a way of thinking that is part and parcel of kind of modern, utilitarian, rationally managed society and saying, okay, well, how do we mathematically lower suffering? Or, alternatively, how do we give this sort of universal moral rights to more and more creatures. And those might be coming from places of good intentions within the kind of Leviathanic way of thinking, but they don't get outside of it.

JOHN ZERZAN: Right, and you wouldn't really think that she would be prey to that, if I can put it that way, all this sort of anthropocentric stuff. And you end up having to rule out so much to maintain your world view, the vegan world view. Fire is no good. Of course, hunting is no good. Anyway, rife with stuff and—maybe further dialogue will, maybe there'll be further changes.

STEVE KIRK: Yeah, I think at this point it may need to be discussed at some length because there is quite a bit of interest in it. And I thought the interview with Layla was really good and kind of revealing in how the argument sort of hit a dead end there. I thought quite obviously. You pointed it out in your closing, but it seemed obvious reading it that it just fell apart. So, I hope people do engage in that.

So, we've talked about two things from *Backwoods 2*, "What Does the World Desire?" and the interview with Layla AbdelRahim, if people are interested.

BELLAMY FITZPATRICK: This is just a commercial. (Laughter)

STEVE KIRK: This is just a commercial.

JOHN ZERZAN: You paid us nicely to publicize it, by the way. (Laughter)

BELLAMY FITZPATRICK: Yeah.

STEVE KIRK: So, you can get your pan-secessionist party off the ground. (Laughter)

BELLAMY FITZPATRICK: Right.

One last comment on the vegan thing, which is, it seems like anyone entertaining that would have to say that we would all have to live in an equatorial or close to equatorial climate and, therefore, basically human migration out of that region was some kind of mistake itself. Which, that's quite a line to draw. It's not incoherent necessarily but it's quite a position.

STEVE KIRK: I think that there has been some anthropological argumentation in that vein, the migration out of Africa, that conquering species and stuff.

Well, veganism, there we go. (Laughter)

As a former vegan it feels like so much cognitive dissonance, all these ideas of veganism that are basically embedded in the writings of Layla and Ria, I think, to a large extent, are these things that I felt like I really had to try and get over at points in my life because they felt so limiting and so inherently civilized. So that's my biggest takeaway.

BELLAMY FITZPATRICK: Yeah, I was vegan for seven years and it's an orientation that's kind of based around hating power, hating power in yourself and being unwilling to countenance your exercising of power because you're sort of withdrawing into this hyper-pacifistic way of being—like, I just don't want to do harm, I just don't want to do harm. And that's—you don't see that in the natural world around you.

STEVE KIRK: Yeah, and how do you even—and then it becomes an issue of quantification like you said earlier.

BELLAMY FITZPATRICK: Sure.

STEVE KIRK: You've got to be able to add it up or else it really doesn't make any sense.

BELLAMY FITZPATRICK: Right, and then you're back in the modern, utilitarian, moral calculus way of looking at things.

JOHN ZERZAN: Well, there's so many topics. I guess you're trying to wrap this up, Steve. I was thinking of the moralism charge and the relationship to values. How one really gets away with calling people moralists if they object to anything. Anyway, it's fascinating to me and I think there's some interesting points on that. But I realize that's a whole other...

STEVE KIRK: Well, we can take five minutes if you want to talk about that. It's fine by me.

BELLAMY FITZPATRICK: Sure.

JOHN ZERZAN: Well, that's my question, let's put it that way—I mean there's so much to say on it, I realize, but some people who go that way pretend there are no values, whereas I think almost everything connects to values. You may try to deny it. So, does that mean you're not a moralist too?

Anyway, the whole thing—it's also, and not to make this even more of a vague thing, but the whole question of ideological contamination or you're an ideologue. That's been thrown at, for example, primitivists. Well, he's just an ideologue—as if their own set of ideas is not as ideological if not more ideological and we don't even know what that actually means. Anyway, all that. I don't know how relevant that is, but it sure intrigues me. And I can think of very recent places this has come up.

For example, if you object to the idea of adults having sex with kids, well what are you? You're some kind of moralizer, you're a moralist, you're a Catholic, you're the Pope—commanding this or that. Well, no, I'm not. But I don't accept that. I just think there's various reasons to really rule it out. But you can't—so, in other words, I'm saying, I don't think you get away with just dismissing with the charge of moralist—then you don't have any values, you're not making these choices which I think are inherently evaluative. Anyway, that whole thing is kind of bugging me. I find it interesting, the ins and outs of that stuff.

STEVE KIRK: Care to comment, Bellamy?

BELLAMY FITZPATRICK: Yeah, sure. Well, I have come around to being a moral realist, I think there really are—there really is such a thing as the objective good and true and beautiful and all those sorts of things. And I think part of this—and this is going to sound hopelessly nerdy or something, but I think it's true—is that I don't think a lot of anarchists in North America at the present moment have really developed a kind of robust ethical theory. And I think that's something that we should do. And I think it's a way for us to communicate with the broader world of people like I was saying before. And it's something I'm going to try to do in my future projects.

And so, I think there's this anarchist way of being that is undeveloped and in some ways sort of adolescent, okay, I'm against everything: I'm against the society and state and capitalism and, therefore, I'm against morality too. And I think what a lot of people really mean is they're against the dominant morality.

And I think what a lot of people are maybe trying to get at is, well I'm very libertarian and I think people should have as much freedom as possible and we shouldn't be imposing certain ways of being on others. And there's a dimension of that that you can take seriously and is positive to a certain degree and that's where, again, I come back to the radical decentralization. So, I would say the charge of moralism, or being ideological, is mostly meaningful when it's directed at this orientation that wants to take over the entire world and impose a certain way of being on everyone. Like I said before, it seems like some anarchists imagine that they're going to impose woke morality on the entire planet and then everyone is going to be happy and I think it's a terrible, ludicrous, hysterical fantasy.

To be an anarchist is to make a moral assertion about how we ought to be and I think we should own that and develop it in a coherent way.

In reference to the pedophilia thing, I assume you're talking about the recent thing where Wolfi [Wolfi Landstreicher] wrote this essay and it boiled up in this latest anarcho-drama, kerfuffle. I fully agree that, as anarchists, we should be against pedophilia and I don't think that Wolfi actually endorses it.

JOHN ZERZAN: Okay, good.

BELLAMY FITZPATRICK: As a friend of Wolfi I feel the need to point that out.

JOHN ZERZAN: Yeah, right, good. I mean, I've known Wolfi since 1980, he's certainly no enemy. We go way back. We certainly disagree on a few things, but I consider him a friend too.

I'm not after Wolfi, it was just an example to some degree and to some degree not. Whether he—I guess he chose not to comment on what he was getting at more precisely, if that would be useful.

But some of these things, you know, everybody sees these things differently. I guess that's obvious enough. But the ITS thing is another case, just to bring in specifics. Individualists Tending Toward the Wild [Editor's note: Also translated Individualists Tending Toward Savagery] and, apparently, I just heard this a couple of days ago, they arrested somebody in Chile who is apparently an ITS person who set off a bomb that injured a postal worker I guess.

I just want to say, I find it just horrific that that's, to some people: well, what's wrong with murdering innocent passersby? I mean, Jesus, have we gone that far down the trail of postmodernism that we can't—to me that's just astounding. And that's why—and this doesn't have to be in the interview I guess—but I have zero respect for people like Aragorn! [Aragorn! of Little Black Cart who recently passed away February 2020, about six months after this interview]. I have a lot of respect for people that, maybe, in a lot of ways, disagree with more. But I just don't see how you pull that up. Oh, it's an interesting source of anarchist ideas? Really? Are you fucking kidding me?

And there was blowback on that, of different kinds, I guess. But when you become a cynical post-modernist to that degree any-fucking-thing goes. And I just find that horrendous, frankly, and I don't mind being called a moralist I guess but, wow, that's just really depressing to think that. And if you consider yourself part of some anti-authoritarian spirit or milieu or what have you, that kind of blows me away.

That's maybe—that's kind of a random thing for me—not random but may it hasn't much to do with what we were talking about.

BELLAMY FITZPATRICK: Well I think it dovetails with the moralism issue. Yeah, I am absolutely not for attacking or harming random passersby. I understand the ITS point of view of saying, well, no one is innocent in this society because we all participate in it. Yeah, that's true in a certain sense, we all countenance hideous things as we go about in our daily life, in some cases, in many cases more or less unconsciously. But the notion of attacking random people because of that is both ethically abhorrent and strategically ridiculous.

And as far as Aragorn!, I consider him a friend, I think both of you know that. I disagree with him about many things—but we also agree on many things. But I'm about— some would say you can't be a free speech absolutist, maybe that's true, but I'm about as close to it as one can be and the—whatever one thinks about ITS and the eco-extremist tendency broadly the treatment that LBC is getting from the milieu for simply publishing a couple of books about ITS—not even their material but material commenting on them—being de-platformed and antagonized because of it

I think is ridiculous. I think the anarchist tendency toward wanting to de-platform or physically attack or otherwise menace people who are simply saying things that some of us don't like is both intellectually cowardly, ethically dangerous, considering the general atmosphere in the United States of cracking down on free speech as a whole. I think it's just a—I wish I could say it were astonishing, it's not astonishing, it's predictably disappointing.

JOHN ZERZAN: Well, what would you say about denying Nazis a platform at, say, the anarchist bookfair? Would you feel like that's an error morally or otherwise?

BELLAMY FITZPATRICK: I would say that an anarchist bookfair is a place of coming together in free association for a particular purpose. And Nazis are obviously statists, authoritarian—I mean Nazi, honestly, at this point, it's a snarl word that has almost no meaning—but if by that we actually mean national socialists, they don't have a place at an anarchist bookfair because they're obviously authoritarian and if they're actually owning their ideas there's no way around that.

I do think, I guess just to throw a different ball into the fray here, that actions such as those by the metropolitan Anarchist Coordinating Council of New York City, where they try to de-platform people on the alt-right by essentially sending repeated whining messages to the authorities of YouTube to take down videos is, again, both ethically objectionable and strategically ridiculous. Because you're literally appealing to authorities with enormous power, far more power than the people on the alt-right who, again, I think are very marginal and unpopular in the mainstream. And calling on these authorities to enhance their censorship and therefore take people out. It's ridiculous. There's nothing anarchist about it.

And it shows a lack of confidence in our actual ideas.

STEVE KIRK: Just to briefly go back though, to what you were saying and how you were sort of talking about the rationalization for possibly not having a national socialist booth at an anarchist book fair. I do feel though, that is precisely the argument that has been given in regard to ITS. I mean, do you see that there are possibly some parallels there? Between the way you rationalize that and the de-platforming of ITS or, more specifically, are you concerned about the de-platforming of LBC as an actual anarchist distributor?

BELLAMY FITZPATRICK: So just let me understand you correctly, are you suggesting that we should be de-platforming Atassa because it is inherently authoritarian, is that what you mean?

STEVE KIRK: No, I actually don't know that I feel that way, but I do think that was the explicit argument of some on the left.

BELLAMY FITZPATRICK: Yeah, and for this set of whiners who apparently have such low confidence in the cognitive and ethical capacities of anarchists—that somehow they will read this book and instantly become so corrupted that we must, therefore, bar them from reading it. Again, it's intellectually cowardly and ITS ought to be interesting to anarchists—I wrote a piece about this in *Black Seed*, I think it was number four—ought to be interesting to anarchists because this is a bunch of former anarchists, people definitely in the deep green strain, who went in a particular direction that I think all three of us agree is wrong, but who came out of the milieu, went into this particular set of tactics and they are informative, both in terms of the how and why they got to the place that they are, and in terms of the reaction of so many within the milieu.

And as I said before, there are many in the North American milieu today who dream of a sudden violent upheaval of the disenfranchised, unfortunate and exploited and if you don't think that that's going to involve awful bloodshed and tremendous privation and, in all likelihood, I think if such a violent uprising were to really happen in an abrupt way, warlordism—all kinds of horror. I think you're deluding yourself. And if you absolutely blanch at this really small, marginal, weak faction killing a handful of people—I don't mean to sound heartless, I do think that killing is wrong—but the people who blanch so much at that I think really need to do some deep searching about the kind of politics that they advocate because they are absolutely countenancing violence, absolutely countenancing so-called innocent people, and I don't say that in a belittling way, dying.

JOHN ZERZAN: Well, of course, there are other ways, there are other possible routes though...

BELLAMY FITZPATRICK: No, I absolutely agree, John, but I'm just saying...

JOHN ZERZAN: ...that doesn't necessarily call for a lot of violence.

BELLAMY FITZPATRICK: I agree with you, John.

JOHN ZERZAN: For instance, '68, ten million people went on strike and there was very little violence. That didn't go very far. Didn't go far enough, long enough, as we know. But the cops and even the army was out-flanked almost immediately so there wasn't much violence. In other words, it's possible that there could be a shift without some traditional blood-bath or something.

BELLAMY FITZPATRICK: I agree with you, John, but the overall record of revolutions doesn't point to that, it points to a lot of the horror that I'm talking about.

JOHN ZERZAN: For sure.

BELLAMY FITZPATRICK: And what you see from these sorts of—big left trinity of Submedia, Crimethinc. [Editor's Note: Crimethinc. has directly addressed this in the piece: "Against the Logic of the Guillotine Why the Paris Commune Burned the Guillotine—and We Should Too"] and It's Going Down, is a lot of this—a lot of imagery and music and that sort of thing that glorifies violent upheaval in a kind of weirdly— weird fusion of old left with new woke, new left. And they have these video images of people riding in pickup trucks with assault rifles and that kind of thing. So, it's like, well, what are you really advocating for.

JOHN ZERZAN: Sure.

BELLAMY FITZPATRICK: And this is the very same bunch of pearl-clutchers who say, oh my god, you killed a few random people this is the worst thing I've ever heard in my life and I just think...

JOHN ZERZAN: Well, I don't quite get how you're putting those two things together. It seems possible that some or possibly all of this ITS is kind of a fraud or a fiction or something. But if they actually did...

BELLAMY FITZPATRICK: I've heard that as well, I don't know.

JOHN ZERZAN: Yeah and, I don't now, if you exalt in it—it's almost as bad. The couple in the park, you kill them and the young woman drunk after a party, she's trying to make a phone call, you go to the phone booth and murder her, and the postal worker who's out behind the post office and he opens the bag and blows himself up. I mean that's—I don't see how you justify that, man. And we can agree...

BELLAMY FITZPATRICK: I'm not justifying it. I'm not justifying it.

JOHN ZERZAN: ...the reason—what you bring in about the left, yeah, of course I think we agree on that. But that's not—I mean we were talking about ITS I thought, and some people made—some people were in bed with that. They were, oh this is interesting and maybe we can make some money publishing this Atassa stuff and that—I just don't think that's very justified. And you don't have to be—I find that offensive. And why not say, well there were people who were once anarchists now they sexually torture a whole bunch of three-month-old babies—you're getting into kind of a stretch. Maybe they didn't kill all that many three-month-old babies so why should we get all worked up about it. That's kind of a twist of logic I would say. It's horrendous or it isn't horrendous, I think it's horrendous.

BELLAMY FITZPATRICK: Let me be clear...

JOHN ZERZAN: And there were a lot of radicals who became fascists in Italy in the '20s, so? I mean, does that make their fascist views interesting, I don't think it does.

BELLAMY FITZPATRICK: Let me be clear, I am not justifying the actions of ITS or any eco-extremists, I am indicting both them and what I see as a hypocritical reaction on the part of certain people who belong to a kind of big anarcho-left who glorify a certain type of hypothetical violence and, at the same time, claim to be absolutely horrified and terrified by the...

JOHN ZERZAN: Right, right, well we all easily agree on that. That's not in question.

BELLAMY FITZPATRICK: And, also, I'm also criticizing the idea of de-platforming and censoring anyone who merely publishes and distributes commentary on the ITS phenomenon. And you brought up the example of, oh, should we be interested in anarchists who became fascists and therefore be interested in that material? Yeah, I actually think that that is valuable to learn from. Why did certain anarchists become fascists? I mean, many people in the alt-right today were formerly anarcho-capitalists or libertarians, how did they go from that to being national socialists?

JOHN ZERZAN: I agree, and I've written about that, too. That wasn't a very good example.

BELLAMY FITZPATRICK: Yeah, Richard Spencer was an anarcho-capitalist.

JOHN ZERZAN: Some people find it kind of fascinating, that these people who didn't mind just murdering people off-handedly. I mean, what's wrong with that? Man, that would be potentially as much of a danger than imagining leftist troops killing everybody in the service of revolution.

BELLAMY FITZPATRICK: Sure.

JOHN ZERZAN: It sounds like even worse. You're deliberately killing people who had nothing to do with anything. Whereas the leftists, you might argue, they have to commit violence against whoever who did this or that. But these victims didn't do anything.

Anyway, I don't want to—I think we probably agree on most all of this, but I don't consider Aragorn! a friend precisely because of this moral vacuous, *anything is fine.*

Hey, we're post-modern, aren't we? It doesn't matter, your story is as good as my story. I don't think they are. I think some stories are more valuable than something else. And there is such a thing as values. And we probably don't disagree about that either, so I'm not trying to pump up a disagreement here, Bellamy.

BELLAMY FITZPATRICK: Sure.

STEVE KIRK: It sounds like we all agree on the broader point about ITS. I think that this issue around de-platforming and trying to draw parallels across different examples to show consistency of some application of some radical idea I don't think that's going to work and we'll probably just go around in circles with that.

Well, probably should wrap...

JOHN ZERZAN: Thank you. And I'm glad to have the chance to talk with you, Bellamy. I really am.

BELLAMY FITZPATRICK: Yeah, very good conversation. Thanks for raising it.